Alexander Mühlberg

Prior-Informed Learning Techniques for Macroscopic and Microscopic Imaging Biomarker Identification

Logos Verlag Berlin

Bibliografische Information der Deutschen Nationalbibliothek

Die Deutsche Nationalbibliothek verzeichnet diese Publikation in der Deutschen Nationalbibliografie; detaillierte bibliografische Daten sind im Internet über http://dnb.d-nb.de abrufbar.

Umschlagabbildung:

Links: Muehlberg, A., Kaergel, R., Katzmann, A., & Suehling, M. (2022). Method for obtaining at least one feature of interest. U.S. Patent No. 11,341,632. Washington, DC: U.S. Patent and Trademark Office.

Mitte: Mühlberg, A. et al. (2023). SEMPAI: a Self-Enhancing Multi-Photon Artificial Intelligence for Prior-Informed Assessment of Muscle Function and Pathology. Advanced Science 10, 220631.

Rechts: Muehlberg, A., Taubmann, O., Katzmann, A. & Suehling, M. (2021). Method for obtaining disease-related clinical information. U.S. Patent Application No. 17/109,332.

ISBN 978-3-8325-5715-7

Logos Verlag Berlin GmbH
Georg-Knorr-Str. 4, Geb. 10,
12681 Berlin, Germany

Tel.:   +49 (0)30 / 42 85 10 90
Fax: +49 (0)30 / 42 85 10 92
http://www.logos-verlag.de

Abbreviations in Alphabetical Order

ACh	Acetylcholine
AI	Artificial Intelligence
AP	Action Potential
APD	Airways-Predominant
AT	Adipose Tissue
AutoML	Automated Machine Learning
BVT	Bias-Variance Trade-Off
COPD	Chronic Obstructive Pulmonary Disease
CSA	Cross-Sectional Area
CT	Computed Tomography
CPH	Cox Proportional Hazards Model
CV	Cross-Validation
DaRe	Data Representation
DL	Deep Learning
(DC)NN	(Deep Convolutional) Neural Network
EML	Extramyocellular Lipids
EPD	Emphysema-Predominant
FAD	Flavin Adenine Dinucleotide
FBP	Filtered Back Projection
GAN	Generative Adversarial Network
GDL	Geometric Deep Learning
GMM	Gaussian Mixture Model
GPU	Graphics Processing Unit
GST	Grey-Value Structure Tensor
HL	Hosmer-Lemeshow Test
IMAT	Intermuscular Adipose Tissue
IML	Intramyocellular Lipids
IR	Iterative Reconstruction
(m)CRC	(Metastatic) Colorectal Cancer
(m)PC	(Metastatic) Pancreatic Cancer
MBT	Institute of Medical Biotechnology
MLS	Muscle-Lipid System
MPM	Multiphoton/Multi-Photon Microscopy
NADH	Nicotinamide Adenine Dinucleotide Hydrogen
NAS	Neural Architecture Search
NMD	Neuromuscular Disorder
OR	Odds Ratio
RECIST	Response Evaluation Criteria In Solid Tumours
Rq.	Requirement
ROC-AUC	Area under the curve of the Receiver Operating Characteristic
ROI	Region Of Interest
SAT	Subcutaneous Adipose Tissue
SEMPAI	Self-Enhancing Multi-Photon Artificial Intelligence
SHG	Second-Harmonic Generation

TML	Traditional Machine Learning, i.e., with handcrafted features
Vanilla	Basic, or uncomplicated version of an algorithm. A starting point, from which more complex variations can be developed
w.r.t.	With Respect To
XAI	Explainable Artificial Intelligence

Conventions

For the sake of readability and as is common in scientific work, the author's own contributions are indicated by the keywords *we/our*, e.g., "*we* engineered the method", while contributions of other authors are explicitly mentioned, e.g., "the setup by author X".

The Oxford (or serial) comma is used throughout this work.

For the sake of clarity, patents and patent applications in the results part are listed with the title of the respective invention disclosure. This aims to enable a better and faster insight into their content. The patent and patent application numbers remain unchanged and provide an unambiguous assignment.

*:= Authors share first/senior authorship

#:= Lead inventor

Figure Index

Table Index

The following text was generated by an artificial intelligence for the prompt: "write me a poem in the style of Rilke about prior knowledge, artificial intelligence, multiphoton microscopy, and computed tomography".

"Artificial intelligence, a wonder of our time,
A force that challenges our prior knowledge,
A tool that helps us to climb
To new heights and to acknowledge.

Computed tomography, a journey through the unseen,
A method that allows us to understand
The inner workings of the body,
And gives us the leading hand.
Multiphoton microscopy, a glimpse into the unseen,
A window into the microscopic realm,
A tool that helps us to glean
The secrets that our eyes overwhelm.

But what of prior knowledge,
Does it fade away like a breeze?
Or does it linger, a foundation,
For all that is yet to be seized?
Rilke wrote of the beauty of mystery,
Of the wonder that lies beyond our sight,
So let us embrace the unknown,
As we delve into the night.
For in the quest for understanding,
We may find that what we know,
Is but a small piece of the puzzle,
And the answers we seek, still to be bestowed.
Let us marvel at the mysteries,
Of artificial intelligence,
Multiphoton microscopy, computed tomography
And all that science brings to existence."

Table of Content

Abstract (English)

Based on the success of artificial intelligence (AI), especially in its application for classification of digital images in the world wide web, its utilisation for automated diagnostics of medical image data has become a major focus. Despite great results with prediction tasks involving big data, a naïve application of deep learning, i.e., use without any prior knowledge about the respective domain, may not be the optimal solution when there are only small amounts of data for the prediction task at hand, which is usually the case in medical studies and biological experiments. Therefore, it may be beneficial to integrate prior information, abbreviated prior, in deep learning or use a more traditional prior-based learning approach.

In this work, **novel macroscopic and microscopic imaging biomarkers for computed tomography (CT) and multiphoton microscopy (MPM) are identified** by developing image processing and learning techniques for biomarker research in **pneumonology, oncology, and muscle research**. From a biological point of view, we were able to improve **fracture discrimination** in CT by detailed analysis of muscle and lipid mixing in the thigh compared to using known bone imaging biomarkers alone. Also, our approach of combining the assessment of macroscopic tumour spread and distribution with machine learning allowed better **survival prediction for cancer patients** than using known biomarkers or naïve applications. We present a calibration method for CT that helped to **preserve the predictive performance of known imaging biomarkers** even when large technical variation and bias induced by the image acquisition process are prevalent. Lastly, we present an AI system that was able to **predict various single muscle fibre properties** from MPM images with better performance than previously known biomarkers.

From a technological point of view, a continuum of learning methods between the old world, i.e., statistics or traditional machine learning with handcrafted features, and the new, i.e., deep learning and meta-learning, is utilised. Our motivation to use prior information leads to novel hybrid learning-based biomarker systems combining prior knowledge with AI. We show that **prior information about a task can effectively be integrated to improve predictive performance of learning algorithms compared to naïve approaches**. Within this work, a prior can be the engineering and integration of suitable features based on biological knowledge about a task, the choice of a suitable data representation for an AI system, or physics knowledge about the nature of noise and artifacts. Our results further indicate that the **representation of the data input to a learning algorithm can be more important than the learning algorithm itself**, and that a suitable data representation for an AI can be different from that for the human observer, e.g., a radiologist. In the final chapter, an **AI is presented that simultaneously determines the required complexity of its neural network architecture, the data representation, and the degree of prior knowledge integration** using meta-learning. It is shown that the integration of priors into deep learning with simultaneous optimisation of the data representation provides the best results, better than naïve deep learning models without priors but also better than models that solely rely on priors.

Altogether, this work presents novel biomarker models for different medical disciplines by combining AI with priors, effectively refining the prior knowledge by AI and regularising complex AI models by priors.

2

Zusammenfassung (Deutsch)

Aufgrund des Erfolgs der künstlichen Intelligenz (KI), insbesondere bei der Klassifizierung digitaler Bilder im World Wide Web, ist ihre Nutzung für die automatisierte Diagnostik medizinischer Bilddaten in den Fokus gerückt. Trotz großartiger Ergebnisse bei Vorhersagen mit großen Datenmengen, ist eine uninformierte Anwendung von Deep Learning, d. h. dessen Verwendung ohne jegliches Vorwissen über den jeweiligen Bereich, möglicherweise nicht die optimale Lösung, wenn nur wenige Daten für die jeweilige Vorhersageaufgabe vorliegen. Dies ist bei medizinischen Studien und biologischen Experimenten in der Regel der Fall. Daher kann es von Vorteil sein, Vorinformationen, auch bekannt als *Priors*, in Deep Learning zu integrieren oder einen klassischeren, auf Vorwissen basierenden Lernansatz zu verwenden.

In dieser Arbeit werden **neue makroskopische und mikroskopische Bildgebungs-Biomarker für die Computertomographie (CT) und Multiphotonenmikroskopie (MPM) identifiziert**, indem Bildverarbeitungs- und Lerntechniken für die Biomarkerforschung in der Pneumologie, Onkologie und Muskelforschung entwickelt werden. Aus biologischer Sicht konnten wir die **Frakturdiskriminierung** mittels CT durch eine detaillierte Analyse der Muskel- und Lipidmischung im Oberschenkel im Vergleich zur alleinigen Verwendung bekannter Knochenbiomarker verbessern. Auch unser Ansatz, die Quantifizierung der makroskopischen Ausbreitung und Verteilung von Tumoren mit maschinellem Lernen zu kombinieren, ermöglichte eine **bessere Überlebensvorhersage für Krebspatienten** als bekannte Biomarker oder uninformierte Anwendungen. Des Weiteren wird eine Kalibrierungsmethode für die CT vorgestellt, die dazu beiträgt, die **Vorhersagekraft bekannter Biomarker zu erhalten**, selbst wenn eine technisch-bedingte Variation und Verzerrung durch den Bildaufnahmeprozess existiert. Schließlich präsentieren wir ein KI-System, das in der Lage war, **verschiedene Eigenschaften einzelner Muskelfasern auf MPM-Bildern mit besserer Leistung als bisher bekannte Biomarker vorherzusagen**. Aus technologischer Sicht wird ein Kontinuum von Lernmethoden zwischen der alten Welt, d.h. Statistik oder dem traditionellen maschinellen Lernen mit „handgefertigten" *Features*, und der neuen Welt, dem Deep Learning und Meta-Learning, genutzt. Unsere Motivation, Vorinformationen zu nutzen, führt zu neuartigen hybriden lernbasierten Biomarker-Systemen, die Vorwissen mit KI kombinieren. Wir zeigen, dass *Priors* **über eine Aufgabe effektiv integriert werden können, um die Vorhersageleistung von Lernalgorithmen im Vergleich zu uninformierten Ansätzen zu verbessern**. Im Rahmen dieser Arbeit kann ein *Prior* die Entwicklung und Integration geeigneter *Features* sein, die auf biologischem Wissen über ein Problem basieren, die Wahl einer geeigneten Datendarstellung für ein KI-System, oder physikalisches Wissen über die Natur von Rauschen und Artefakten. Unsere Ergebnisse deuten außerdem darauf hin, dass die **Repräsentation der Daten, die in einen Lernalgorithmus eingegeben werden, wichtiger sein kann als der Lernalgorithmus selbst**, und dass eine geeignete Datenrepräsentation für eine KI eine andere sein kann als für den menschlichen Betrachter, z.B. einen Radiologen. Im letzten Kapitel wird eine **KI vorgestellt, die die erforderliche Komplexität ihrer neuronalen Netzwerkarchitektur, die Datenrepräsentation und den Grad der Integration von *Priors* mit Hilfe von Meta-Learning selbst bestimmt**. Es wird gezeigt, dass die Integration von *Priors* in Deep Learning bei gleichzeitiger Optimierung der Datenrepräsentation die besten Ergebnisse liefert, besser als uninformierte Deep Learning-Modelle, aber auch besser als Modelle, die ausschließlich auf *Priors* basieren.

Diese Arbeit präsentiert innovative Biomarker-Modelle für verschiedene medizinische Fachbereiche, die KI und Vorwissen miteinander kombinieren. Einerseits wird hierbei das Vorwissen durch Einsatz von KI verbessert, andererseits werden komplexe KI-Modelle durch das Vorwissen regularisiert.

1. Introduction and State-of-the-Art

1.1. Macro- and Microscopic Imaging Technologies for Biomarker Identification

1.1.1. Computed Tomography (CT) for Macroscopic Imaging

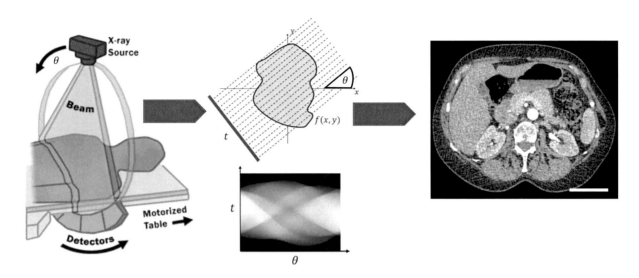

Figure 1. Schematic drawings of CT image acquisition. Information about the attenuation of photons by the object is collected in the detectors (left) and used to reconstruct the object f(x,y) (centre, top) slice-wise from its sinograms (centre, bottom), i.e., the attenuation profile with offset t from the detector origin for all rotation angles θ. For simplification, parallel beams were assumed in the central image. An exemplary axial CT image of the abdomen is shown on the right. Bar: 10cm. Modified from U.S. Food & Drug Administration[1] and Van Aarle et al.[2].

Computed tomography (CT) is an x-ray-based diagnostic imaging modality. CT utilises that the radiodensity of biological tissue, i.e., the relative inability of electromagnetic radiation to pass through it, is dependent on the material's physical density and its mass attenuation coefficient[3]. The latter is a function of the atomic number of the material and the photon energy. The attenuation of x-ray photons can thus be used to determine the radiodensity and, accordingly, also tissue distribution and tissue properties within objects. As shown in (Fig. 1., left), a CT gantry, consisting of a photon emitter and a detector, rotates around an object, and the attenuation of x-ray photons after passing the object is measured in the detector. By rotating the gantry, attenuation profiles (projections) for different gantry angles can be obtained. The projection information over all angles is termed sinogram (Fig. 1, centre bottom). The operation that enables the generation of a sinogram from a 2D image is known as Radon transform[4], and the inverse Radon transform, accordingly, is the required operation to solve the CT image reconstruction problem. Applying image reconstruction on a sinogram yields an axial 2D image (Fig. 1, right), and by movement of the motorised table, a 3D image, i.e., volume, of the object can be generated. The radiodensity shown in CT images is usually provided in the Hounsfield scale, in which distilled water at standard pressure has a Hounsfield unit (HU) of 0 and air of -1000 HU.

A common reconstruction algorithm is the filtered back projection (FBP) which is a practical and fast solution for the ill-posed inverse Radon transform problem[5]. FBP can be derived by the central slice theorem that links 1D projections and the corresponding axial 2D image by Fourier and inverse Fourier transforms, thereby enabling reconstruction of 2D images from the sinogram in a simple and fast process[3]: the projection for each angle (of the sinogram) is convolved with a reconstruction kernel, and the results are integrated over all angles in 2D. The reconstruction kernel is selected specifically for the radiological examination at hand as a trade-off between noise- and detail-level, with the hard kernels providing sharper images with more details, and soft kernels providing images with less noise and artifacts[3]. More advanced algorithms for image reconstruction in CT are based on iterative reconstruction (IR) which involves iteratively solving an optimisation problem to estimate the image from the projection data. This method results in improved image quality, but it also requires a longer reconstruction time compared to FBP. IR minimises the difference between the measured projection data and the estimated projections based on the current estimate of the image, improving the image estimate with each iteration[6,7]. IR can incorporate prior information, such as model-based knowledge of anatomy or structure, to improve the image estimate, reduce noise and artifacts, and speed up convergence. Besides the reconstruction algorithm, further important influencing factors for the image formation are the tube voltage which determines the energy spectrum of the photons, and the tube current which determines the photon quantity. Higher tube current results in lower statistical noise as more photons arrive at the detector, but it also leads to higher radiation exposure. Higher tube voltage is needed if the object would otherwise attenuate the photons too much, but this also results in lower image contrast.

CT is by far the most widely used tomographic imaging modality with approximately 300 million scans per year worldwide compared to 95 million magnetic resonance imaging (MRI) scans[8]. It is commonly the imaging method of choice in oncology, pneumonology, in the emergency department, and for the assessment of bone structures or calcium deposits. Compared to MRI, its more frequent application in oncology is based on the lower cost and wider availability of CT while yielding sufficient diagnostic information for most cancer types. In pneumonology, it is based on the better imaging capabilities to assess slight density differences of air-filled organs, like the lungs. In the emergency room, especially the significantly faster image acquisition compared to MRI is the deciding factor for CT. Finally, CT is superior for the assessment of bone structures or calcium deposits, e.g., in the coronaries, as a result of its imaging mechanism based on attenuation.

Variants of CT are used in different clinical and research scenarios. Contrast medium is injected in around 40% of all CT examinations[8] to increase the contrast of structures of interest thus enabling, e.g., the indirect assessment of tumour metabolism[9,10] or the diagnosis of cardiovascular diseases[11]. Dual-energy CT[12] uses photons from two energy spectra to exploit the high energy dependence of the mass attenuation coefficient of materials with high atomic numbers. It, thereby, allows material decomposition to assess kidney stones, gout, or iodine content and even enables the generation of synthetic images like virtual non-contrast, i.e., the contrast medium is virtually removed from the images. Current research and also an emerging clinical application field is photon-counting CT in which the detector measures single photons and their energy (energy-resolving) instead of the sum of all energies of the photons reaching the detector (energy-integrating)[13]. In photon-counting CT, commonly four energy bins are used for the incoming photons, and the number of photons in each of these bins is counted. This allows material decomposition, in analogy to dual-energy, in each scan and yields even more multi-spectral information. Furthermore, this technique provides increased contrast-to-noise ratio (CNR) and spatial resolution. It can reduce noise and various artifacts, like beam hardening, and enables to discriminate photons with energies below the lowest energy bin, resulting in an elimination of electronic noise. To enable photon-counting CT, a novel detector was constructed which replaces the common scintillation detector by a semi-conductor detector that directly transforms photons into electric pulses according to their energy. Lastly, for preclinical research, micro-CT can be employed, which offers higher resolution and is often used for *in vitro* and *in vivo* small animal imaging[14].

1.1.2. Label-free Multiphoton Microscopy (MPM) for Microscopic Imaging

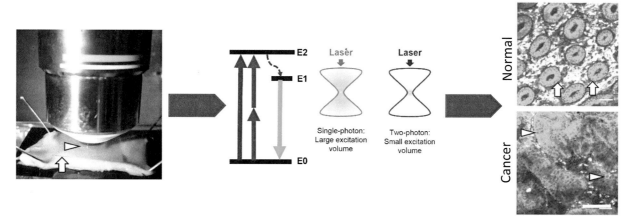

Figure 2. Schematic drawings of label-free MPM image acquisition. A colorectal tissue sample is examined by MPM (left) for oncological diagnostics. The sample is covered by a coverslip (arrow) and a water drop (arrowhead) is used as immersion. (centre) Two photons are absorbed to excite a fluorophore. When the electron falls back to the ground state, fluorophore-specific light Is emitted. The excitation volume for two-photon microscopy is smaller than for single-photon applications due to excitation based on the simultaneous absorption. (right) MPM enables to differentiate normal colon tissue from cancer. Arrows and arrowheads indicate the epithelial cells of normal and cancer tissue, respectively. Images were acquired at 780nm excitation and using bandpass filters of 417/60nm (blue: epithelial cells and fibrous tissue), 480/40nm (green: immune cells and epithelial cells), and 629/56nm (red: immune cells). Bar: 100µm. Modified from Matsui et al.[15] and Kreiss et al.[16].

Multiphoton microscopy (MPM) is an advanced optical microscopy technology that utilises a pulsed laser as excitation source[17]. In this approach, two laser photons of lower energy are simultaneously absorbed for the excitation of a single fluorophore. This fluorophore can either be an exogeneous marker that is added to the sample by fluorescent staining[18] or a native molecule in the sample. When the excited electron falls back to its ground state, light is emitted from the fluorophore, which can then be recorded. One main advantage of this technique over conventional fluorescence microscopy is that the excitation is naturally confocal and without simultaneous light emission from out-of-focus planes. This is due to the constraint of spatially-confined simultaneous photon absorption. In addition, since the probability of scattering increases for photons of higher energy, MPM manages to produce images deeper in the tissue by using photons of lower energy (Fig. 2).

MPM also allows second-harmonic generation (SHG) imaging by using short laser pulses[18]. Samples with second order susceptibility properties are polarised by the electric field component of the incident light. This polarisation can be described in a Taylor series. If the electrical field is strong enough, the otherwise negligible second term of this series induces a secondary wave at exactly twice the frequency. In comparison to two photon-induced fluorescence, SHG photons have twice the energy as incident photons[19,20]. Therefore, the SHG signal can not only be separated from light of the source but also from fluorescence by using appropriate dichroic mirrors and filters. The application of SHG, however, is restricted to biomolecules without inversion symmetry such as tubulin, collagen-I, and myosin-II[21]. As a main advantage, SHG is sensitive to the orientation of the examined probes. This can, for example, be used to analyse muscle fibres, more specifically the structural state of myofibrillar myosin polymers[22].

Most optical microscopy methods use exogenous markers for the examined samples. However, these can interfere with the probe's biology or its intrinsic binding homeostasis. Label-free approaches utilise the non-linear excitation of native fluorophores, e.g., FAD or NADH, in combination with the SHG of native biomolecules. These label-free variants are particularly interesting as they enable the use of MPM for endoscopic examinations, i.e., a so-called multiphoton endomicroscopy[23].

1.2. A Short Tale of Finding Patterns in Data: From Statistics to Neural Architecture Search

Statistics is defined as the *"science of collecting, analyzing, presenting, and interpreting data"*[24]. The first statistical approaches of collecting data for interpretation date back to the antique and were censuses in Egypt and China. The first mathematical methods for statistical inference, i.e., deduction of general patterns from the data, were then developed in the Arab world around 800 AD[25]. The modern field of statistics emerged at the end of the 19th century and the beginning of the 20th century[26]. Modern statistics can be divided in descriptive statistics which measures properties of the data like standard deviation or mean, and inferential statistics which assumes that the analysed data are sampled from a larger unknown population and tries to draw conclusions about this population, for example by estimators and hypothesis tests. Inferential statistics uses mathematically derived rules and heuristics to determine the optimal so-called bias-variance-trade-off (BVT) for construction of multivariable models (Fig. 3, left). The BVT describes that complex models can learn subtleties of the data to infer properties of interest which we term labels in the course of this work, but are also prone to statistical noise or random patterns, causing the models to not generalise well[27]. Also, when fitting the model, removing samples can cause large changes, or variance, of the model. This is called a high variance model or overfitting. The process of reducing variance is termed regularisation. At the other extreme are very simple models, e.g., linear regression with one feature. These models are not as complex as required, therefore, they have a bias regarding the description of the real patterns in the data. This is called high bias models or underfitting. Statistics assumes the form of the model function and commonly uses linear models. The model complexity is, therefore, mostly determined by the number of used parameters/features in the model, and regularisation is accordingly achieved by reduction of the feature set size. For that purpose, heuristics for dimensionality reduction, such as multiple testing correction or best subset selection, can be used to determine the optimal BVT.

The term machine learning was formulated by Arthur Samuel in 1959[28]. While the field of artificial intelligence (AI) was initially more concerned with symbolic approaches and expert systems, the data-driven approach of AI (Fig. 3, right), machine learning (ML), began to flourish in the 1990s.

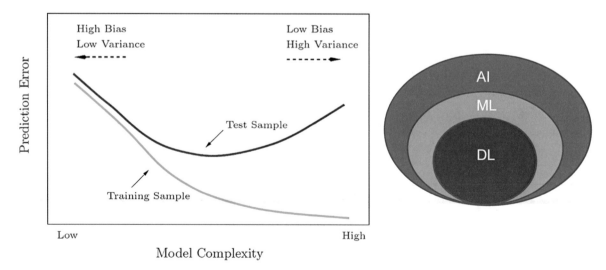

Figure 3. Bias-variance trade-off (left) and sub-disciplines within AI research (right). Left figure modified from Hastie et al.[27].

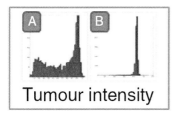

Tumour intensity

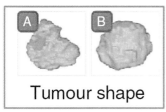

Tumour shape

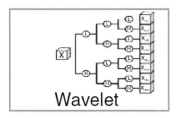

Tumour texture

Wavelet

Figure 4. Handcrafted radiomics features of four classes. Taken from Aerts et al.[30].

The goal of ML is to achieve the best possible prediction for a label. To distinguish ML from inferential statistics, we can refer to the definition of Bzdok et al.[29]: *"Inference creates a mathematical model of the data generation process to formalize understanding or test a hypothesis about how the system behaves. Prediction aims at forecasting unobserved outcomes or future behavior".* Accordingly, while statistics uses the whole data set and estimates the generalisability of a (linear) statistical model from one set and by model assumptions, the generalisability of ML models, which are usually non-linear, is estimated by testing them on unseen data. For this purpose, the data are split into training, development (dev) and test set. By applying the trained models to the dev set, the ML model and its complexity can be adapted, both through the choice of a method, such as random forest, support vector machine, or a neural network (NN) with few layers, and through the choice of model-specific parameters, so-called hyperparameters. The models determined in this way are evaluated for the test set. The BVT definition can be adapted for ML: a high variance model explains the training set well but shows bad predictive performance for the test set (overfitting), while a high bias model already performs badly on the training set (underfitting) but not necessarily worse on the test set (Fig. 3, left). A common evaluation scheme is N-fold cross-validation, where in N-iterations disjunct parts of the data are removed. After N-iterations, each sample was removed once. The remaining data in each iteration are used as training set for the ML algorithm, while the removed data are used as test-set. The results for the N different test sets are then micro- or macro-averaged.

In the past, ML used mainly features as input. These features are handcrafted and either arbitrary or based on intuition about problem. We term ML systems with handcrafted features traditional machine learning (TML) in the course of this work. Exemplary applications of TML in medical research are genomics and, especially for oncological imaging, the field of radiomics[30,31]. In the latter, a large number of arbitrary intensity, texture, and shape features (Fig. 4) are extracted from target lesions, and a TML model is trained to predict a label, e.g., the survival time of a patient. A breakthrough for radiomics was the work by Aerts et al., in which they discovered a four-feature signature (Aerts signature), consisting of three texture and one shape feature, that improved survival prediction with TML compared to the state-of-the-art radiological assessment in independent clinical cancer cohorts[30].

With increasing digitisation and the success of the world wide web, big (volumes of) data became available, enabling models with increased complexity. Neural networks with at least one hidden layer and a non-linear activation function are universal function approximators[32], i.e., they can learn every (continuous) function. Furthermore, they can be made arbitrarily complex by adding more layers[a]. Neural networks with many layers are called deep neural networks, and the according branch of research deep learning (DL, Fig. 3, right). A NN is composed of layers of neurons. Each neuron is a function parametrised by its parameters, i.e., weights and a bias, which are adapted during learning in the backpropagation step. For this, a loss (function) has to be defined, which is a metric to quantify the deviation between NN

---

[a]Theoretically, this can also be achieved by using many neurons in just one layer. In practice, this approach does not work.

output and ground truth of the labels. In the backpropagation, partial derivatives of the loss w.r.t. the NN parameters are calculated and passed on to an optimiser. The optimiser updates the NN parameters according to the gradient. The iterative minimisation of the loss is termed learning or training. A number of techniques had to be developed specifically for DL. For example, it was discovered that the utilisation of the Rectified Linear Unit (ReLu) activation function enables deeper networks by preventing vanishing gradients. Another breakthrough in ML for image data was achieved by the so-called deep convolutional neural networks (DCNNs). These make the crafting of features superfluous, but instead they automatically generate feature maps by convolutions which are applied to an image. The convolution kernels, which can also be called filters or feature detectors, consist of weights that are adapted in the backpropagation. The introduction of max-pooling layers for DCNN[33] and the efficient parallelised training on graphics processor units (GPUs) finally led to the breakthrough of DL in scientific and industrial practice by beating many image classification benchmarks in 2012, most prominently for ImageNet[34]. Additionally, techniques like batch normalisation or dropout are often used to regularise the training of DL models.

The training of a DCNN for the automated classification of handwritten digits is shown in Fig. 5. An input image with the handwritten digit "2" is entered. Through iterative convolutions with a 5x5 kernel and max-pooling, feature maps are created and flattened to yield ten output neurons after the fully connected layers. A softmax activation is applied to the ten outputs, resulting in a probability for the corresponding digit. The cross-entropy loss is used to quantify the deviation from the ground truth.

Major problems of DCNNs are their lack of explainability but also their instability. For example, a changed pixel in the image of a stop sign by an adversarial attack can cause a complete misclassification by the DCNN. Such instability is dangerous in the medical field and thus a large obstacle for the large-scale introduction of AI in hospitals. Further scepticism is generated by the lack of human comprehension about the decision process of DL[35]. Therefore, the branch of decision explanation or more generally Explainable AI (XAI) is becoming increasingly important also in order to detect decisions of DCNNs that are based on spurious correlations in the training data (*Clever Hans* effect[36]). Many XAI techniques, such as LRP saliency maps[37], were already developed to prevent DL from being a black box by highlighting the image regions (pixels, voxels) that are relevant for the DL decision process. Furthermore, the concept of known operator learning[38] might improve the stability of DL algorithms and, therefore, potentially decrease the susceptibility to adversarial attacks. In this approach, known operators, e.g., FBP layers with learnable parameters in the case of CT reconstruction, are integrated into DCNNs. In this way, prior knowledge can effectively be utilised in DL, and it could be shown, besides its success for a large number of practical examples, that this methodology generally reduces the maximum error bound of a model[38], i.e., the expected error for unseen data. Prior knowledge about a target function is an efficient way to reduce the bias and variance of a model simultaneously[39,40] by semantic regularisation. In this work, we call models without any utilisation of prior knowledge naïve.

Another cutting-edge branch of research is automated machine learning (AutoML) or meta-learning. In this research field, ML systems are automatically identified for prediction tasks (*"learning to learn"*). On the downside, especially AutoML for NN architectures, the so-called neural architecture search (NAS)[41], is extremely computation-intensive as it has to train and evaluate a large number of NNs. However, NAS offers, in addition to being a more systematic approach for finding suitable NN architectures, the advantage that it requires less human effort and thus also brings economic benefits for industry.

Finally, we need to introduce the term data representation (DaRe) for the comprehension of later sections. DaRe in the context of AI research is the representation of the data that are passed on to the learning algorithm. Although DCNN should learn appropriate feature maps independently through its application of convolutions to the images, in practice, an appropriate DaRe is important. For example, a 2D DaRe of a tumour may be more appropriate than a 3D DaRe for prediction tasks with smaller amounts of data, as the probability of overfitting is reduced. This can be sub-summarised under *curse of dimensionality*[42], i.e., a larger input data dimensionality needs more training data. As a rule of thumb, there should be at least five training examples for each dimension in the representation[43] although this requirement may be weakened for DCNNs[44]. Other examples of DaRe modification are pre-processing of images, e.g., denoising, or, in the case of TML, feature selection. In current practice, the DaRe of an image that is fed into a learning method is often arbitrary. Different techniques for a DCNN, e.g., the max-pooling layer, also internally reduce the dimensionality. Additionally, there are explicit AI methods for data dimensionality reduction, such as sparse autoencoders[45], where the data must pass through a bottleneck, leaving only essential image traits.

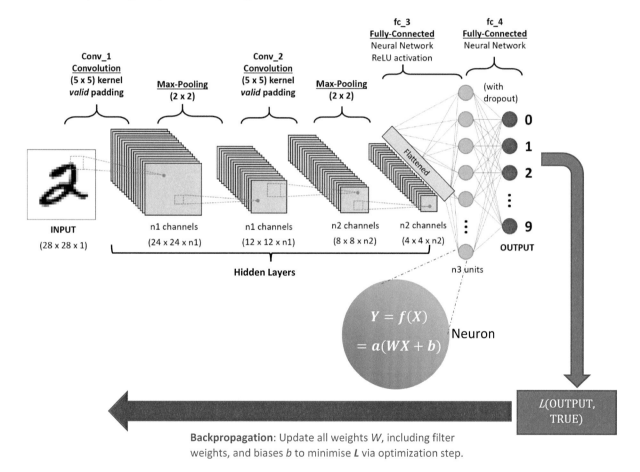

Figure 5. DCNN example for classification of handwritten digits. The image of the digit or a batch of images is fed into the DCNN. Convolutions with learnable kernels are applied on the input to yield feature maps, followed by max-pooling layers. The feature maps are flattened and fed into the first fully connected layer of neurons, with each neuron being a function f(X), described by weight matrix W and bias b. An activation function a, in this case ReLu, and dropout for regularisation are applied on the outputs of the pre-final layer. Those are fully connected to the final layer. The activation function of the output neurons, i.e., the outputs of the final layer, is a softmax, where the sum of all outputs from the neurons always equals 1. An output of 0.8 of the first output neuron means a correct classification of the digit as a "2". The deviation between output and ground truth is quantified by the loss L which is the cross-entropy in this case. The optimiser for the parameter update is stochastic gradient descent. For simplicity, just one neuron was shown exemplarily in more detail. Modified from Saha et al.[46].

1.3. Medical Background

1.3.1. Myology

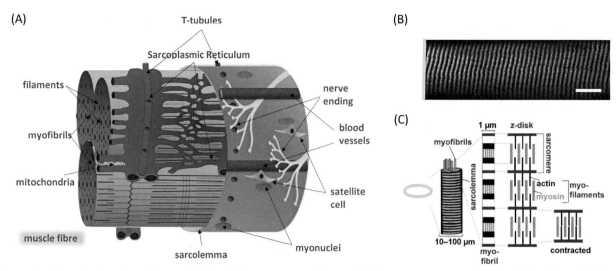

Figure 6. Schematic drawings of a muscle fibre. (A) Composition of a muscle fibre, its nerve connection and blood supply (B) Image of a muscle fibre in a mouse model via SHG. (C) Muscle fibre in relaxed and contracted state. To link each subfigure, a single fibre is indicated by yellow. Bar: 25μm. Modified from Mukund et al.[47] and Schmidt-Nielsen et al.[48].

Skeletal muscles, or in this work simply muscles, enable us to mechanically interact with the physical world. A reduction in muscle function initially makes it harder to perform complex activities, such as lifting or running, but in advanced stages may lead to complete immobility and inability to participate in life. Muscle tissue has a hierarchical and regular structure with a predominantly parallel arrangement of muscle fibres, myofibrils, and myofilaments in sarcomeres flanked by Z-discs (Fig. 6). Force is generated by interaction of actin and myosin within the sarcomeres[49]. This process is dependent on Ca^{2+} ions and ATP[50]. When a sufficiently strong depolarisation from the nerves reaches the axon hillock of a motor neuron, an action potential (AP) is initiated. The AP leads to the release of acetylcholine (ACh) at the neuromuscular junction[51]. The ACh induces a postsynaptic excitatory potential at the sarcolemma which, if a potential threshold is exceeded, travels as a muscular AP along the T-tubules to the sarcoplasmic reticulum, where Ca^{2+} is released. Those Ca^{2+}-ions lead to a conformational change of troponin and tropomyosin molecules, thereby enabling actin and myosin to interact[52]. ATP is consumed during this interaction. Muscle dysfunction can be categorised as either being part of an age-related decline in muscle mass and function (sarcopenia) or as part of the group of neuromuscular disorders (NMD) which affect different parts of the motor unit, i.e., muscle, nerves, and neuromuscular junctions[53]. NMDs that originate from within the muscle itself are referred to as primary myopathies. One primary myopathy is Duchenne muscular dystrophy that is caused by a genetically induced absence of dystrophin[54]. Dystrophin protects muscle fibres, especially their sarcolemma, from mechanical damage by contraction that prevents excess Ca^{2+} from entering the cell. Its absence initiates a pathophysiology cascade which is not perfectly understood[54], causing fibres to become necrotic and be replaced by connective tissue and adipocytes. This progressive disorder affects about 1 in 3,500 boys[55].

Microscopically, single muscle fibres can be examined by MPM. Those fulfil conditions of high degree of organisation and orientation, which allows also the use of SHG. The SHG signal commonly originates from endogenous proteins of interest and is dependent of their molecular orientation and assembly. For striated muscle, additionally, the state of actomyosin and myofibrils can be obtained since the polarisation dependencies of SHG signals change between relaxed and rigor state (cf. section **1.1.2**).

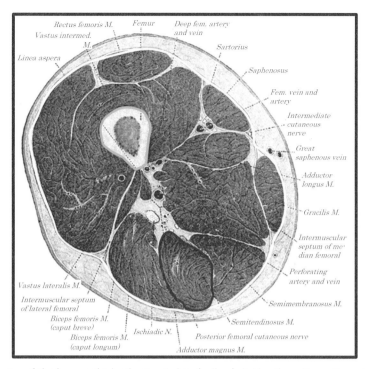

Figure 7. Schematic drawing of the human thigh. The *Fascia lata* (yellow) divides the adipose tissue (light grey, outside bone) into SAT and IMAT. IMAT which is deposited within the anatomical muscles, e.g., the *Semitendinosus* muscle (purple), is termed EML. The macroscopic muscle then consists of muscle tissue and EML. Bone (green) interacts both directly with muscle tissue and indirectly with adipose tissue. Modified from Mühlberg et al.[56].

Macroscopically, muscles are commonly examined with MRI, the method of choice in the clinical routine due to the higher soft tissue contrast compared to CT (cf. section **1.1.1**). Macroscopic imaging also allows to assess the environment of muscle. According to the *Mechanostat* theory, muscle and bone form a unit, and the force of the muscle induces growth of the bone[57]. To describe the study situation regarding the interaction of muscle tissue, adipocytes, and lipids, compartments must be introduced, exemplarily for the thigh (Fig. 7): the anatomical muscles of the thigh are enclosed by a thin layer of connective tissue, the *Fascia lata*, which divides the adipose tissue of the thigh into subcutaneous adipose tissue (outside, SAT) and intermuscular adipose tissue (inside, IMAT). Elevated IMAT level is associated with various pathologies, such as muscle weakness[58] and metabolic disorders[59]. IMAT can also displace muscle tissue and be deposited directly within anatomical muscle borders. These IMAT deposits are called extramyocellular lipids (EML) which in turn can be of diagnostic interest and are, for example, associated with decreased muscle strength[60]. However, lipids can also be deposited as droplets close to the mitochondria of muscle fibres, i.e., within the muscle tissue, and are then called intramyocellular lipids (IML)[61]. The exact biochemical interaction of lipids and specifically adipocytes with muscle tissue in this regard is not yet fully understood, although attempts of explanation exist. In addition, there is also an indirect interaction between bone and adipose tissue: adipose tissue protects bone structures against falls by cushioning[62].

Although muscles are mostly examined macroscopically with MRI, CT provides, besides a higher resolution, also the possibility to quantify the so-called muscle density[63] which is the mean CT value (HU) of muscle tissue. Originally, it was assumed that this measurement quantifies IML, but it was shown that it correlates only weakly with lipid content in magnetic resonance spectroscopy[64]. At present, it is assumed that, in addition to lipid content, muscle fibre density and protein degradation are also reflected[63,65]. The muscle density is associated with both decreased muscle strength[66] and the tendency of bone fracture[67] in the elderly.

1.3.2. Oncology

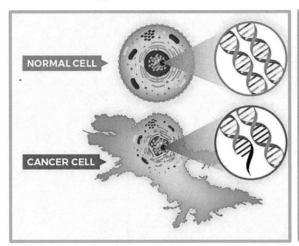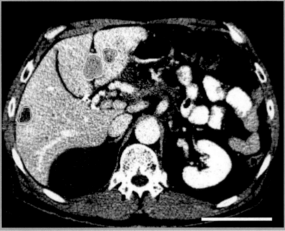

Figure 8. Scheme of carcinogenesis and liver metastases in CT. (left) Cancer is a genetic disease which is induced by DNA damage, resulting in uncontrolled growth of the cancer cells. (right) Liver metastases from primary pancreatic cancer. Bar: 10cm. Modified from National Cancer Institute[68] and Gebauer et al.[69].

The clinical field of oncology treats cancer patients. *"Cancer is a genetic disease in which certain body cells grow uncontrollably and spread to other parts of the organism, which can occur almost anywhere in the human body"[68]*. The process of cell division, known as mitosis, allows for the formation of new cells when necessary. Meanwhile, the process of apoptosis ensures the removal of damaged or older cells from the organism. However, when cellular regulatory mechanisms are compromised, abnormal or damaged cells may continue to divide and form masses of metabolically active tissue, referred to as tumours. These tumours can be classified as benign (noncancerous) or malignant (cancerous). Malignant cancer cells can, in addition to proliferation, also invade healthy tissue and deposit in other (distant) organs to form metastases. The main cause of cancer is DNA damage, that occurs naturally or is induced by external factors such as smoking or alcohol intake[68].

In 2020, more than 19.3 million new cancer cases were reported worldwide and 10.0 million cancer deaths, which makes it the second leading cause of death[70]. There are more than 100 types of cancer, therefore, we focus on colorectal cancer (CRC) as an example. Around 50% of all CRC patients develop liver metastases, 15-20% lung, and 1-4% brain metastases[68]. This is termed metastatic colorectal cancer (mCRC).

CRC is most commonly monitored via CT imaging, with intervals of 8 to 12 weeks between subsequent scans[71]. For prognosis with CT, oncologic TNM staging[72] is used, a set of semi-quantitative metrics with which the radiologist assesses the size of the primary tumour (T), lymph node involvement (N), and (distant) metastases (M). Typical outcome variables to be estimated are survival time, either continuous or according to established thresholds, e.g., 1-year survival, and therapy response. The latter is assessed by Response Evaluation Criteria In Solid Tumours (RECIST)[71]. In RECIST, not the whole tumour burden is assessed quantitatively but only target lesions. In studies, continuous survival time is commonly modelled using survival regression, a statistical method that takes into account censored data from patients whose survival time is not known past the censoring point[73].

A current promising research topic is the identification of genes that impact a variety of different cancer types, which is termed pan-cancer research[74]. By this approach, first shared patterns of genetic mutations could be identified, that occurred a long time before oncogenesis[75]. In the future, pan-cancer analyses might enable a very early screening for cancer.

1.3.3. Pneumonology

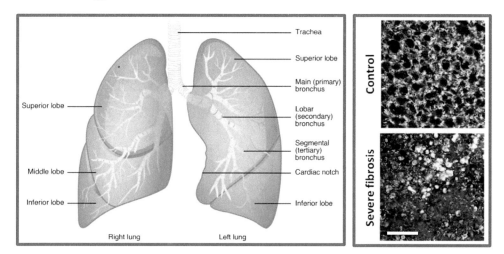

Figure 9. Macroscopic and microscopic structure of the lungs. (left) Scheme of the lung anatomy. (right) Healthy and fibrotic lung tissue in a mouse model revealed by MPM. Signals of NADH (415-485 nm, green), collagen (395-415 nm, blue) and FAD (540-580 nm, red) are shown. Yellow: overlay of NADH and FAD signals. Bar: 50µm. Modified from Betts et al.[76] and Kreiss et al.[16].

The lungs are the organs responsible for enrichment of the blood with oxygen and release of carbon dioxide. Each lung is divided into lobes (three on the right, two on the left) and traversed by bronchi, i.e., the airways. These branch out from the main bronchus, through the lobar bronchi, the segmental bronchi to the terminal and respiratory bronchioles (Fig. 9 left, bronchioles are not resolved). In the lung space around the bronchial tree, the lung consists of lung parenchyma. This parenchyma is composed of thin-walled alveoli, which are responsible for gas exchange of oxygen and carbon dioxide between respiratory bronchioles and the blood.

In times of a pandemic triggered by COVID-19, research in the field of pneumonology gained momentum. In COVID-19, the SARS-CoV-2 virus invades alveolar cells via the ACE-2 receptor[77]. While severe outcomes have been specifically triggered by the inflammatory processes[78] in the lungs leading to potential lung failure, a long-term effect is fibrosis of the lung parenchyma, i.e., replacement of alveoli by connective tissue[79]. An exemplary MPM image of lung fibrosis in a mouse model is shown in Fig. 9, right.

A longer known lung disease, Chronic Obstructive Pulmonary Disorder (COPD) is the third leading cause of death and one of the most prevalent diseases worldwide, with estimated 7.6% (high-income countries: 9.3%) of world population suffering from the disease[80] in the year 2019. In the parenchyma, different forms of emphysema (panlobular, paraseptal, centrilobular) appear, i.e., air-filled spaces that replace collapsed alveoli. Clinically, emphysema can be detected by low-attenuation areas (LAA) on CT scans, which indicate air-filled spaces. In COPD, also the airways are affected, e.g., in small airways disease the terminal bronchioles are obstructed, and the air from the trachea cannot reach the respiratory bronchioles. This can be detected in CT either directly by thickened peripheral airways or indirectly, e.g., by a mosaic attenuation pattern of the parenchyma. COPD is defined by irreversible airflow obstruction and limitation, which can be measured by spirometry.

CT is the standard imaging modality in clinical routine, especially due to the better capability to image air-filled organs compared to MRI, and enables an assessment of phenotypes, e.g., emphysema-predominant (EPD) or airways-predominant (APD) COPD. Microscopic methods applied for lung biopsies can support the diagnosis but are mainly used in research (Fig. 9, right).

1.4. Objectives

The main objective of this work was to establish different variants of prior-informed learning systems to identify novel **macroscopic and microscopic quantitative imaging biomarker models in myology, oncology, and pneumonology, thereby fostering precision medicine**. For this purpose, we developed a variety of image processing pipelines, which include segmentation, registration, feature engineering, and standardisation techniques for biomedical images. In a second step, we engineered learning methods for the features or the images directly to identify novel biomarker models.

A second objective was to evaluate the integration of prior information (prior) into the learning method. Thus, in the course of this work, we utilise physics priors about the imaging process (section **2.2.1**), or biological priors, such as known diagnostic criteria or intuition about a problem (sections **2.2.2, 2.3**), and combine those priors with the learning process or data-driven optimisation to tackle prediction tasks. We compare complex but naïve methods, e.g., DL without priors, with simple statistical and state-of-the-art methods that use priors.

In the upcoming sections, results of our work regarding above-mentioned branches are presented.

2. Results and Discussion

2.1. Macro- and Microscopic Pneumonological Imaging Biomarkers

2.1.1. Macroscopic: Unravelling the Interplay of Image Formation, Data Representation and Learning in CT-based COPD-Phenotyping Automation

(1) **Mühlberg, A.,** Kärgel, R., Katzmann, A., Durlak, F., Allard, P. E., Faivre, J. B., … Rémy-Jardin, M.*& Taubmann, O.* (2021). Unraveling the interplay of image formation, data representation and learning in CT-based COPD phenotyping automation: The need for a meta-strategy. *Medical Physics*, *48*(9), pp. 5179-5191.

(2) **Muehlberg, A.#,** Taubmann, O., Katzmann, A., Denzinger, F., Lades, F., Kaergel, R., … & Suehling, M. (2022). A Radiological Intuition-guided Automated Deep Learning System. *U.S. Patent Application No. 17/382,588.*

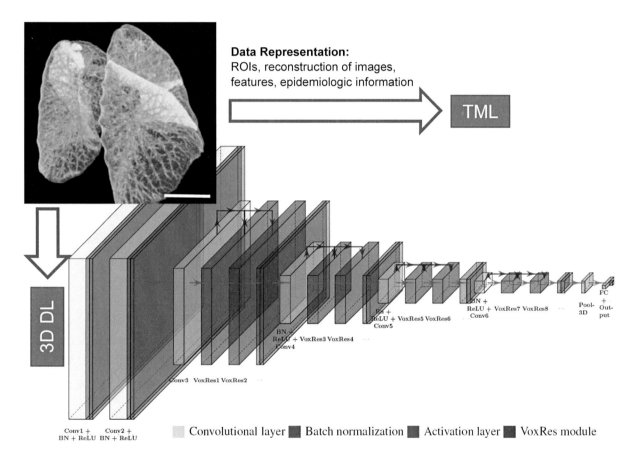

Figure 10. Setup to compare different TML algorithms and varying DaRe (top) with 3D DL (bottom). A variety of TML algorithms, including AutoML, was integrated to compare the importance of DaRe with the importance of the TML algorithm and 3D DL. Bar: 10cm. Modified from Mühlberg et al.[81].

For COPD phenotyping with CT, a large body of ML research exists that aims to automatically detect or classify one or a few phenotypes. These phenotypes include different forms of emphysema (EPD) but also phenotypes of the class airways diseases (APD). In the respective publications, mainly new DL algorithms are presented, most often for studies with a small sample size and arbitrary data representation (DaRe) and study design. The Hôpital Calmette of the University Centre of Lille provided us with a COPD imaging study of 981 patients, in which radiologists patient-wise labelled 13 different phenotypes such as different emphysema types, air trapping, or visible small airways disease.

Our idea in publication **(1)** was to compare the importance of the learning algorithm, including AutoML[82] for the classification of phenotypes, with the importance of the DaRe (cf. section **1.2**). The choice regarding DaRe, for example, includes the granularity of analysed regions of interest (ROIs), e.g., compartmentalisation of the lung in different lobes vs. analysing the lung as a whole, or the additional utilisation of ROIs in other organs. The impact of DaRe as a result of different reconstruction kernels was assessed. This parameter is usually selected by radiologists as part of the study design and impacts the image formation (cf. section **1.1.1**). Also, the decision to exclude or impute features of airways generations too small to measure in some patients was evaluated. Finally, the impact of the study design decision to integrate epidemiological features of age and sex for phenotyping was analysed.

A second motivation was to present a study that can be considered as a baseline of expected predictive performance with adequate sample size for a variety of phenotypes, with which novel research in the field of DL can benchmark their developed algorithms. For that, TML was combined with only one holistic set of COPD-specific handcrafted features for all phenotypes but with a large data set of 981 patients.

Automated lung lobe[83], airways[84], and organ segmentation[85] algorithms were integrated. A set of-COPD-specific features was extracted from the lobes and airways, and radiomics features were extracted from further ROIs in organs. DaRe was varied, as explained above, and used with different TML algorithms, and the prediction performance for each phenotype was protocolled. Multivariate statistics was employed for these meta-data to simultaneously quantify the importance of DaRe and learning algorithm on predictive performance for each phenotype. We additionally compared the results of phenotype classification using optimised TML and DaRe to the results obtained using a 3D DL model, i.e., naïve DL without optimisation of DaRe. For this purpose, a VoxResNet[86,87] was trained with the entire lung parenchyma as input for an automated classification of emphysema. The DL algorithm was pre-trained on COPDGene[88], a cohort with large sample size of several thousand, where it achieved convincing predictive performance.

The results show:

- The choice of the learning algorithm was not as important as the choice of the DaRe.
- Advanced learning techniques like 3D DL or AutoML yielded inferior automation performance for non-optimal DaRe in comparison to less complex TML techniques with optimised DaRe.
- TML with a holistic set of COPD-specific features achieved good predictive performance for most phenotyping tasks, especially for EPD phenotypes. For EPD, the performance was comparable to DL algorithms specifically engineered for single phenotypes that are reported in the literature.
- An AI system has different requirements regarding DaRe than a human reader, e.g., TML shows a better predictive performance with a soft reconstruction kernel for classification of APD phenotypes, which is in contrast to the radiologist's clinical routine for this task. The prior knowledge of the radiologist about the optimal DaRe can, therefore, be misleading and cannot simply be transferred to the ML setting.
- Radiologists should systematically evaluate and adapt the DaRe and further meta-parameters defined in the study design to optimise the performance of AI algorithms.

The results led us to the concept of automatically optimising the DaRe together with the DL architecture, which is described in the filed patent **(2)**. In this, a DaRe optimisation is combined with a NAS algorithm and a toolbox for radiologists to collect potentially useful DaRes based on their intuition, which are then propagated to all samples of the patient cohort. From this DaRe collection, the system selects the most promising while simultaneously searching for the optimal NN architecture. In 2023, a student will explore and implement this concept in the course of a Master's thesis.

2.1.2. Microscopic: Lung Fibrosis Grading Automation by Integrated MPM-Raman Data

(3) Kreiß, L.*, Ganzleben, I.*, **Mühlberg, A.,** Ritter, P., Schneidereit, D., Becker, C., ... & Waldner, M. <u>Label-free analysis of inflammatory tissue remodeling in murine lung tissue based on multiphoton microscopy, Raman spectroscopy and machine learning.</u> *Journal of Biophotonics*, *15*(9), e202200073.

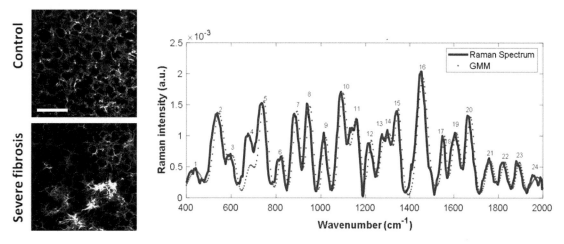

Figure 11. Integrated MPM-Raman data consisting of 3D MPM image (left, SHG shown here) and Raman spectrum with and without GMM fit (right). Bar: 50µm. Modified from Kreiss et al.[16].

The idea of author Kreiß was to engineer an advanced measurement setup which provides integrated data by label-free MPM, including SHG, and Raman spectroscopy of a sample. The added value of this setup was evaluated for automated lung fibrosis diagnosis in a mouse model (Fig. 11). We used the generated multi-modal data with TML to predict the semi-quantitative Ashcroft score[89] for histological assessment of lung fibrosis and a derived binary class (healthy: Ashcroft score <2/fibrosis: else).

For the MPM data, three established features were extracted, the dispersion parameter k, cellularity, and SHG density. For Raman spectra, our idea was to reduce the dimensionality by a Gaussian mixture model (GMM) fit to the spectra. This is possible, as the number of expected intensity peaks in the spectra is known, i.e., prior information. The GMM reduces the dimensionality of a sample from 523 spectral data points down to 72 GMM features (width, height, and centre of each peak) (Fig. 11, right). This allows a proper statistical analysis with multiple testing correction and our hypothesis was that the inclusion of these Raman-GMM features improves TML prediction in comparison to using the whole spectrum (cf. *curse of dimensionality*, section **1.2**). Since the prediction of Ashcroft score is an ordinal regression task, for which TML modelling is not trivial, TML models for multi-class classification and also regression were trained. A TML approach was additionally employed for the binary classification healthy/fibrosis. Different feature signatures were evaluated, and the predictive performance of the TML models was compared with the best MPM feature, cellularity. Additionally, we visualised the potential of multi-modal data for knowledge discovery by hierarchical clustering.

The results show:

- For a simple binary classification healthy/fibrosis, TML with multi-modal datasets yielded no added value, since MPM-based cellularity already provided strong predictive performance.
- For the more difficult task of predicting the Ashcroft score, the combination of TML and integrated Raman-GMM and MPM datasets improved predictive performance compared to uni-modal analyses.
- The dimensionality reduction by GMM enhanced Ashcroft score prediction.

2.2. Technically-invariant Macroscopic Oncological Imaging Biomarkers

This section addresses the influence of technical variation and bias on the identification and utilisation of biomarkers. Both in field radiomics, i.e., TML, and in DL, so-called batch effects[90] can be introduced by the measurement process. An example is a tumour that is classified by an AI in CT as benign with soft reconstruction kernel and as malignant with hard reconstruction kernel (cf. **section 1.1.1**). The AI, i.e., TML or DL, cannot distinguish between the more heterogeneous texture of the tumour based on biological properties and the heterogeneity due to technical properties, i.e., higher noise level. This problem is commonly present when using the texture of a ROI in TML and DL models. Methods need to be developed that can generate technically-invariant imaging biomarkers and systems.

2.2.1. Development of a Predictive Internal Calibration: Technome & DeepTechnome

(4) **Mühlberg, A.,** Katzmann, A., Heinemann, V., Kärgel, R., Wels, M., Taubmann, O., ... Nörenberg, D.* & Rémy-Jardin, M.* (2020). The Technome-a predictive internal calibration approach for quantitative imaging biomarker research. *Scientific Reports, 10*(1), 1103.

(5) Langer, S., Taubmann, O., Denzinger, F., Maier, A., & **Mühlberg, A.** (2022). DeepTechnome: Mitigating Unknown Bias in Deep Learning Based Assessment of CT Images. *arXiv Preprint arXiv:2205.13297.* **(Accepted, Bildverarbeitung für die Medizin 2023)**

(6) **Muehlberg, A.#,** Kaergel, R., Wels, M., & Suehling, M. (2021). A Method for Technically-Invariant Quantitative Imaging Biomarkers. *U.S. Patent No. 10,922,813.* Washington, DC: U.S. Patent and Trademark Office.

(7) **Muehlberg, A.#,** Kaergel, R., Katzmann, A., & Suehling, M. (2022). Feature-Enhanced Computed Tomography for Technical Invariant Imaging Biomarkers. *U.S. Patent No. 11,341,632.* Washington, DC: U.S. Patent and Trademark Office.

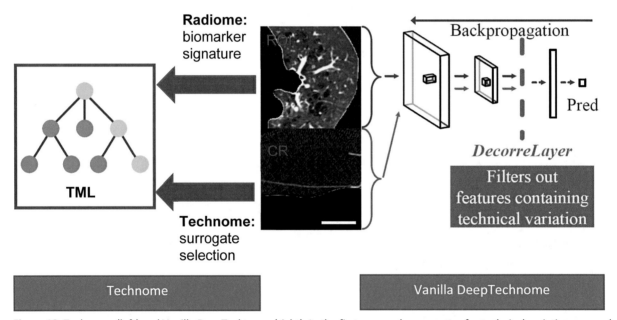

Figure 12. Technome (left) and Vanilla DeepTechnome (right). In the first approach, surrogates for technical variation are qualified and selected by the technome from a CR to calibrate the radiome, i.e., a biomarker signature extracted from a ROI, for a TML approach. In the latter approach, a DCNN is calibrated against technical variation by filtering out neurons of the feature maps that are correlated with the same features in a CR. Bar: 5cm. Modified from Langer et al.[91].

Some mean value measurements with CT, such as bone mineral density (BMD) in the field of musculoskeletal imaging, need to be calibrated by in-scan phantoms to ensure, to a certain degree, technically-invariant measurements[92]. This is required, as the HU values of bone are strongly dependent on the photon energy (cf. section **1.1.1**). There are also approaches that calibrate BMD measurements phantomless by using the patient's adipose tissue for a so-called internal calibration[93].

Our idea was to develop an internal calibration for texture measurements used in TML and DL models, as texture is highly dependent on a variety of technical characteristics[94], including noise and artifact level. All methods in this section use control regions (CRs) in the image that are supposed to capture characteristics of technical influences, e.g., the air in Fig. 12.

In the granted patent **(6)** and the publication **(4)**, surrogates for technical variation, e.g., representative features for noise, are extracted from CRs and used to calibrate TML-based models. We introduce the technome, a kind of hub that captures technical influences on biomarker signatures and selects surrogates for calibration. Since it is not clear which surrogates are suitable, and since selection of the best surrogates used for internal calibration also depends on which ROI features/biomarkers need to be calibrated, a qualification for these surrogates is required (i.e., selection rules) (Fig. 12, left).

Qualification criteria for surrogates were defined and implemented, which can be interpreted as physics priors, i.e., knowledge about the nature of noise. These physics priors are based on information acquired from statistics of the clinical cohort (*in vivo* qualification), phantom tests (*in vitro* qualification), and simulations (*in silico* qualification).

Yet, a calibration of a feature can rarely be perfect, therefore the technome has two different operational modes with distinct objectives: The stabilisation mode aims to keep a feature as invariant as possible to technical influences. The predictive mode, however, aims at combining the calibration with a good predictive performance of a TML model with the calibrated feature signature.

In the predictive mode, a combined loss is minimised, which is the sum of a calibration loss and a training loss. The calibration loss enforces the highest possible qualification of selected surrogates. By minimising the combined loss, the qualification of the surrogates and the training of a TML model for the prediction of a biological label are achieved simultaneously. Since it is not clear how the different qualification criteria have to be weighed against each other in the calibration loss, free parameters were introduced as weights for each criterion and determined while minimising the combined loss. This determination in the course of the optimisation yields insights into how helpful the respective qualification criterion is for the construction of a good predictive model. The combined loss is minimised by Bayesian optimisation. In predictive mode, qualified surrogates are integrated in the TML model in addition to the biomarker signature which is thereby calibrated.

We applied the method to the established Aerts signature[30] and a cohort of 686 mCRC tumours to predict 1-year survival. As a second proof-of-concept was required, we applied it to a low-attenuation area (LAA) signature for the lung parenchyma, consisting of two known imaging biomarkers %LAA-910 and %LAA-950[88], and a cohort of 676 COPD patients to classify centrilobular emphysema. Technical variation in the mCRC cohort is large and mainly a result of varying scan and reconstruction parameters between patient scans, while it is lower in the COPD cohort and based on different noise and artifact levels as a result of varying patient attenuation characteristics and a hard reconstruction kernel.

To benchmark our method, we used TML with (i) the respective signature without technome calibration, (ii) the signature and all possible (non-qualified) surrogates in the CRs that were selected by a TML feature selection method, and (iii) the signature calibrated by a variant similar to the RAVEL[95] approach from MRI research. Finally, (iv) naïve radiomics for the mCRC cohort and TML with (v) the LAA signature and soft reconstruction kernel for the COPD cohort were evaluated. We varied the sample size to compare the usefulness of the method in experimental studies with small amounts of data, which is further elaborated in the publication **(4)**.

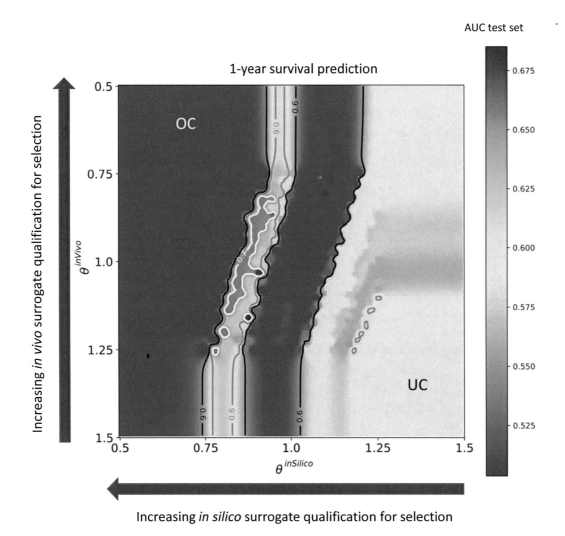

Figure 13. Association between the qualification of surrogates integrated in a TML biomarker model and the model's generalisability, as measured by its ROC-AUC on a test set and presented by contour plots. The signature to be calibrated consists of the four Aerts features, the prediction task is 1-year survival. Lower qualification weights, i.e., axis values, indicate higher required qualification for surrogates to be selected. Integrating highly qualified surrogates, as determined by both *in vivo* and *in silico* qualification, leads to a highest generalisability of the TML models. However, setting the qualification criteria too high results in underfitting as no surrogates are selected for the model. On the other hand, setting the criteria too low results in overfitting, as too many surrogates are selected for the model. This results in overconstrained (OC) or underconstrained (UC) models, respectively. The results can be interpreted as an indication or even proof that the *in vivo* and *in silico* qualification criteria really calibrate a biomarker signature. They helped to reproduce the predictive performance of the Aerts signature, which it has shown before in a large number of studies for cancer outcome prognosis. Note: a model from the "narrow mountain" of high predictive performance was identified by minimising the combined loss for the training data. Modified from Mühlberg et al.[96].

The results show that:

- Both prediction tasks benefitted from technome calibration. mCRC 1-year-survival was predicted by the calibrated Aerts signature with a ROC-AUC of 0.67, compared to (i) 0.50 (TML: linear model) or 0.58 (TML: random forest) without it. Similarly, centrilobular emphysema classification improved from (i) 0.61 (TML: linear model) or 0.58 (TML: random forest) to 0.82.
- The models with predictive internal calibration were better than (ii) the naïve approach and (iii) the RAVEL-like calibration in both cases.
- The calibrated Aerts signature yielded considerably better 1-year-survival prediction compared to (iv) a naïve radiomics approach.
- Centrilobular emphysema classification with hard reconstruction kernel and technome calibration performed better than (v) with soft kernel without calibration. The soft reconstruction kernel is usually preferred in clinical research for LAA measurements, providing images with lower noise level.
- Suitable surrogates were best identified by their association (correlation) with features of the biomarker signature in clinical cohorts (*in vivo* qualification) and by the reproducibility of this association with simulations (*in silico* qualification) (Fig. 13).
- Interestingly, the technome calibration of the Aerts signature yielded only slightly worse predictive performance for mCRC 1-year survival than an elaborate DL architecture with sparse auto-encoder (0.67 vs. 0.71) as proposed by Katzmann et al. **(14)** (cf. section **2.2.3**).

Although the technome approach can calibrate known biomarker signatures or models consisting of few features, thus improving their reproducibility and predictive performance, it cannot be used for exploration with a large number of arbitrary features or with DL. Since DL is expected to scale better with big data and is, thus, crucial for an industrial application, we worked on a concept to transfer the technome to DL. In the granted patent **(7)** we suggest a setup where a DCNN applied on the ROI (radiome), e.g., a tumour, and a DCNN applied on a CR (DeepTechnome) do interact. The radiome evolves to identify novel biomarkers while the DeepTechnome uses qualification criteria for the CR to generate its feature maps: If DeepTechnome feature maps correlate with those of the radiome and fulfil further qualification criteria, the radiome is penalised by an additional loss enforcing the feature maps of the radiome to evolve orthogonally to technical features of the CRs and, thus, become technically-invariant.

In a first vanilla (cf. abbreviations section) approach **(5)** of this concept, author Langer implemented a variant in the course of his Bachelor's thesis, in which the evolution of the radiome's feature maps is influenced by the DeepTechnome in a specific layer of the DCNN, the DecorreLayer (Fig. 12, right): The feature maps of the radiome are also extracted in a CR and if features of ROI and CR are correlated (cf. *in vivo* qualification), the corresponding feature in the radiome is filtered out by the DecorreLayer, effectively debiasing the radiome from unknown bias.

This implementation was tested for two scenarios. In the first, an artificial bias was introduced by imbalanced sampling of images acquired with hard kernel and those with soft kernel from the COPD data for DCNN training: for the emphysema group of the training set, more samples were drawn with soft kernel, and for the group without emphysema, more samples with hard kernel. In the second scenario, synthetic noise was added to the training data with emphysema to introduce an artificial bias for DCNN training. In both cases, we did not modify the test set. It was compared how well DeepTechnome could remove the bias from the training set for predicting the test set. In both scenarios, DeepTechnome improved predictive performance but only for smaller NN architectures.

2.2.2. Trajectomics: Analysing Macroscopic Growth Patterns of Cancer

(8) **Mühlberg, A.,** Holch, J. W., Heinemann, V., Huber, T., Moltz, J., Maurus, S., ... & Nörenberg, D. (2021). The relevance of CT-based geometric and radiomics analysis of whole liver tumor burden to predict survival of patients with metastatic colorectal cancer. *European Radiology, 31*(2), pp. 834-846.

(9) Gebauer, L.*, Moltz, J. H.*, **Mühlberg, A.,** Holch, J. W., Huber, T., Enke, J., ... & Maurus, S. (2021). Quantitative Imaging Biomarkers of the Whole Liver Tumor Burden Improve Survival Prediction in Metastatic Pancreatic Cancer. *Cancers, 13*(22), 5732.

(10) Rist, L., Taubmann, O., **Mühlberg, A.,** Denzinger, F., Thamm, F., Sühling, M., Nörenberg, D., Holch,J., Maurus, S., Gebauer, L. , Huber, T. & Maier, A. Spatial Lesion Graphs: Analyzing Liver Metastases with Geometric Deep Learning for Cancer Survival Regression. **(Accepted, *2023 IEEE International Symposium on Biomedical Imaging*)**

(11) **Muehlberg, A.#,** Katzmann, A., Durlak, F., & Suehling, M. (2021). Dispersion-based Tumor Analytics System. *U.S. Patent Application No. 17/009,954.*

(12) **Muehlberg, A.#,** Taubmann, O., Katzmann, A., & Suehling, M. (2021). Geometric Deep Learning-based Whole Tumor Burden Analytics System. *U.S. Patent Application No. 17/109,332.*

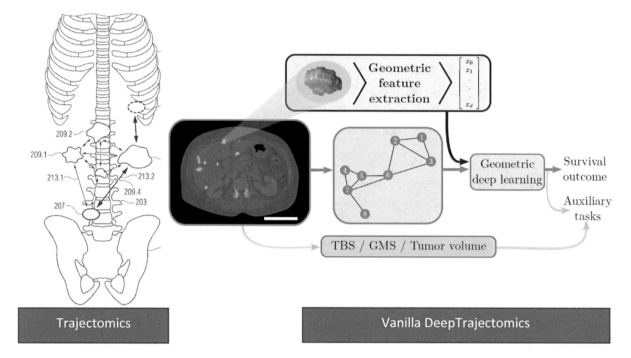

Figure 14. Trajectomics (left) and DeepTrajectomics (right). In the trajectomics approach, the spread and distribution patterns of lesions are used with statistics or TML for automated oncological diagnosis. In the vanilla version of DeepTrajectomics, geometric DL is employed in which lesion properties, such as shape or volume, are described as nodes and inter-lesion properties as edges of a graph. Prior-based features are used as auxiliary tasks. Bar: 10cm. Modified from Mühlberg et al.[97] and **(10)**.

In this section, we present an alternative approach to yield technically-invariant biomarkers. Our idea was to develop TML/DL systems that do not use texture, thus, minimising the impact of technical variation. Instead, we utilise the prior knowledge from TNM that distant metastases are associated with worse survival outcome[72]. We translate this to organ-level and introduce the Geometric Metastatic Spread (GMS), i.e., a feature signature describing the maximum spread and distribution of lesions in an organ. In contrast to RECIST, radiomics, and DL approaches, we analyse all lesions within an organ, i.e., the whole tumour burden, and their distribution and not just single target lesions.

This concept is described in **(8)** and **(11).** While in the filed patent **(11),** we additionally describe how the GMS can be integrated into DL, including its utilisation as auxiliary tasks, in publication **(8)**, we use the GMS with TML models to predict survival time and 1-year survival of mCRC patients.

We introduce the term trajectomics for (non-linear) model building with geometric features that quantify temporal snapshots of tumour trajectories (Fig. 14, left), such as the GMS, in analogy to the branches of genomics or radiomics that use genes or image features of target lesions as features. In publication **(8)**, we compare trajectomics with (i) the established Aerts signature[30] and (ii), a rather novel imaging biomarker for patient survival, the tumour burden score (TBS)[98]. The TBS is the Pythagorean addition of the number of lesions and the diameter of the largest lesion. In addition, we compare trajectomics with (iii) a model of known imaging biomarkers, such as tumour volume or primary tumour sidedness, and (iv) a model of clinical biomarkers, such as histology or TNM staging.

A pipeline consisting of an automated whole tumour burden segmentation of the liver[99] and feature extraction, especially geometric features of the distribution, was applied to 103 patients with mCRC. For this cohort, models (i)-(iv) were evaluated in the statistics setting as well as with TML for classification (1-year survival) as well as for survival time prediction.

The results show:

- The univariate biomarker (ii) TBS performed best in the statistics setting, i.e., for linear models.
- In the TML setting, i.e., for non-linear complex models, trajectomics and (i) Aerts signature were superior to (ii) TBS, (iii) the imaging biomarker model, and (iv) the clinical model for 1-year-survival prediction. Trajectomics was the best model for survival time prediction.
- Although (i) the Aerts signature was slightly superior to the trajectomics model for classification, the latter does not use texture and is, therefore, presumably technically invariant.

Trajectomics was used a second time in the publication of Gebauer et al. **(9)**. In this study, most GMS features for liver lesions were significantly associated with survival time of metastatic Pancreatic Cancer (mPC) patients but not used in the statistical multivariable model after feature selection. TML was not utilised in this publication.

Based on the initial results of trajectomics with TML and inspired by the advent of geometric deep learning (GDL)[100], a combination of both appeared promising. We, therefore, developed a concept **(12)** which uses GDL to combine geometric analyses of the whole tumour burden with properties of the individual lesions. We term this approach DeepTrajectomics. Here, the relationships of tumours to each other are described by edges of a graph, i.e., the graph topology, while properties of the lesions, such as volume or shape (but theoretically also texture), are described by nodes of a graph.

With a first vanilla implementation within the course of a Master's thesis, author Rist evaluated this concept for mPC survival time prediction **(10)** (Fig. 14, right). He implemented and compared different GDL architectures and graph topologies and found that a GDL approach was superior to a number of models, including TML with the Aerts signature and DCNNs. Also, the impact of priors GMS, TBS, and tumour volume as auxiliary tasks as proposed in **(11)** was shown to stabilise the training. Of the GDL architectures, a differential pooling architecture provided the best predictive performance and enabled to effectively incorporate more node features, while a graph attention network still outperformed all other non-GDL benchmarks. Texture features of the nodes did not yield any additional benefit.

Currently, a publication is in preparation that explores shared trajectomics patterns of mPC and mCRC to establish trajectomics in the field of pan-cancer research[101] (cf. section **1.3.2**).

2.2.3. Further Contributions to Deep Learning and Oncological Biomarker Research

(13) Katzmann, A., **Mühlberg, A.,** Sühling, M., Nörenberg, D., Holch, J. W., & Groß, H.-M. (2018) TumorEncode-Deep Convolutional Autoencoder for Computed Tomography Tumor Treatment Assessment. *2018 International Joint Conference on Neural Networks (IJCNN)*, pp. 1-8, IEEE.

(14) Katzmann, A., **Muehlberg, A.,** Sühling, M., Noerenberg, D., Holch, J. W., Heinemann, V., & Groß, H.-M. (2018). Predicting lesion growth and patient survival in colorectal cancer patients using deep neural networks. *1st International Conference on Medical Imaging with Deep Learning (MIDL).*

(15) Katzmann, A., **Mühlberg, A.,** Sühling, M., Nörenberg, D., & Groß, H. M. (2019). Deep Metamemory-A Generic Framework for Stabilized One-Shot Confidence Estimation in Deep Neural Networks and its Application on Colorectal Cancer Liver Metastases Growth Prediction. *2019 IEEE 16th International Symposium on Biomedical Imaging (ISBI)*, pp. 1298-1302, IEEE.

(16) Katzmann, A., **Muehlberg, A.,** Suehling, M., Nörenberg, D., Holch, J.W., & Groß, H.-M. (2020). Deep random forests for small sample size prediction with medical imaging data. *2020 IEEE 17th International Symposium on Biomedical Imaging (ISBI)*, pp. 1543-1547, IEEE.

(17) Katzmann A, **Mühlberg A**, Sühling M, Nörenberg D, Maurus S, Holch J.W... & Groß, H.M. (2019). Computed Tomography Image-Based Deep Survival Regression for Metastatic Colorectal Cancer Using a Non-proportional Hazards Model. *International Workshop on PRedictive Intelligence In Medicine*, pp. 73-80, Springer.

(18) Katzmann, A., Taubmann, O., Ahmad, S., **Mühlberg, A.,** Sühling, M., & Groß, H. M. (2021). Explaining clinical decision support systems in medical imaging using cycle-consistent activation maximization. *Neurocomputing, 458*, pp. 141-156.

(19) Wels, M. G., Lades, F., **Muehlberg, A.,** & Suehling, M. (2019). General purpose radiomics for multi-modal clinical research. *Medical Imaging 2019: Computer-Aided Diagnosis, 10950*, pp. 1047-1054, SPIE.

(20) Woźnicki, P., Westhoff, N., Huber, T., Riffel, P., Froelich, M. F., ... **Muehlberg. A.,** ... & Nörenberg, D. (2020). Multiparametric MRI for prostate cancer characterization: Combined use of radiomics model with PI-RADS and clinical parameters. *Cancers, 12*(7), 1767.

(21) Enke, J. S.*, Moltz, J. H.*, D'Anastasi, M., Kunz, W. G., Schmidt, C., ... **Mühlberg, A., ...** & Hahn, H.(2022). Radiomics Features of the Spleen as Surrogates for CT-Based Lymphoma Diagnosis and Subtype Differentiation. *Cancers, 14*(3), 713.

(22) **Muehlberg, A.#,** Taubmann, O., Katzmann, A., Durlak, F., Wels, M., Lades, F., ... & Suehling, M. (2022). Radiomics Imaging Biomarker Assistant (RIBA) - Research Companion for Clinical Imaging Studies with Novel (and Established) Technologies. *U.S. Patent Application No. 17/462,140.*

(23) Katzmann, A.#, Ahmad, S., Suehling, M., & **Muehlberg, A.** (2022). Cycle-consistent Generative Adversarial System for Deep Decision Explanation. *U.S. Patent Application No. 17/476,630.*

(24) Katzmann A.#, Kratzke L., **Muehlberg A.,** & Suehling M. (2022). Sparse Lung Nodule Characterization for Differential Diagnosis from CT images. *U.S. Patent No. 11,138,731.* Washington, DC: U.S. Patent and Trademark Office.

This section presents conceptual contributions in the field of DL and oncological imaging biomarker research. Unlike the main theme of this work, these do not specifically address prior-informed learning. Instead, the first part of this paragraph presents general contributions to novel DL methodology especially for small data, while the second part discusses contributions to conventional radiomics approaches for biomarker identification in clinical studies.

For the publications of Katzmann et al. **(13-18)**, in which the author of this work contributed mostly as a second author, there was a close conceptual exchange. However, the implementation was without exception carried out by author Katzmann.

In publication **(13)**, the application of sparse autoencoders[45] for the prediction of tumour growth, i.e., the treatment response, is presented. The motivation was to yield a DL algorithm that generalises for smaller sample size and is less affected by technical variation. An autoencoder is an unsupervised DL approach that compresses DaRes and preserves only essential information to reconstruct the input,

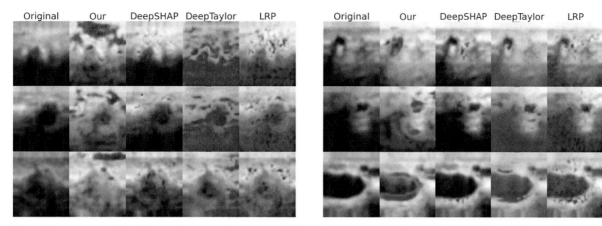

Figure 15. Examples of decision explanations for benign (left) and malignant (right) lung lesions. Indications for malignant (red) and benign (blue) lesions, as quantified by each algorithm, are depicted as a coloured overlay. Taken from Katzmann et al.[102].

effectively reducing the dimensionality. Additionally, the method needed only single tumour slices of two treatment time points and used a smaller, less complex DCNN architecture. As a major advantage, the needed input of just one segmented slice per time point means only slightly more effort for a radiologist compared to the clinical routine assessment of RECIST by the largest diameter of the target lesion(s) (RECIST-LLD). The results show that this approach can yield better predictive performance than RECIST-LLD. A histogram equalisation was applied on the slices to modify the DaRe, which enabled a more effective training of the autoencoder by reducing the data variance. It could also be shown that a decision explanation based on saliency maps was linked to biologically relevant patterns. In publication **(14)**, the used time points were harmonised by now ensuring that the first time point used is always before treatment start. Furthermore, 1-year-survival was introduced as an additional prediction task and for evaluation, a 4-fold CV was employed instead of just one split. The method was compared with radiomics and RECIST-LLD and showed significantly better predictive performance than both for various classification metrics. This concept was extended for an application to lung tumours by additional utilisation of semi-synthetic training data generated by a generative adversarial network (GAN)[103] and filed as a patent **(23)**.

In publication **(15)**, a novel DL method that combines confidence estimation with a variant of curriculum learning[104], i.e., the sampling of data based on the difficulty to learn, is presented. Confidence estimation for a prediction is particularly important for medical data. The presented method internally estimates its uncertainty and accordingly adjusts its data sampling strategy. Thus, samples that are harder to learn are presented to the DCNN more frequently than simple ones. This is achieved by passing hints about the ground truth label to the DCNN in the training phase. Samples that are more dependent of hints are harder to learn, i.e., lower confidence is given, and are, accordingly, sampled more frequently. The approach was compared to a baseline DL approach for the benchmark data set CIFAR-10 and radiological data and was slightly superior to both.

In publication **(16)**, another attempt is made to tackle the problems of DL with small data. For this purpose, bootstrapping aggregating (bagging), which is known from the TML method random forest[105], is employed. Random forests consist of ensembles of decision trees for bootstrapped subsets of the data. By bagging, i.e., the aggregation of classifiers trained on bootstrapped subsets, the method is regularised. In the proposed method, instead of using handcrafted features as input to the random forest as in the standard approach, a DCNN is trained, and the outputs of the neurons of the pre-final NN layer are used for the decision trees of each bootstrapped subset, effectively combining bagging with DL. Since the training of this method was very slow, our idea was to speed it up by using weight inheritance known from NAS research. The method was used again for mCRC 1-year-survival prediction with decent results.

In publication **(17)**, a DL variant for survival regression to predict survival time is presented (cf. sections **1.3.2, 2.2.2**). Survival regression is most often based on the Cox proportional hazards model (CPH)[73]. This is an established statistical model, where the hazard probability is fitted as a multiplication of a time-dependent base hazard probability with a feature-dependent exponential term. CPH relies on some (distributional) model assumptions and additionally requires handcrafted features. Katzmann et al. present a variant where the hazard function is estimated by a DCNN with the advantage that fewer model assumptions need to be fulfilled and no crafting of features is required. The method was evaluated on three datasets, including two cancer datasets, where it showed decent performance.

In publication **(18)**, the problem of missing explainability of DL systems is addressed by a new decision explanation method. Most decision explanation methods to visualise the importance of input pixels/voxels show lower quality when the amount of data is low. Katzmann et al. address this problem by developing a decision visualisation method that utilises a cycle-consistent generative adversarial network (CycleGAN)[106] model for activation maximisation. The developed method was semi-quantitatively compared in a user study for the CIFAR-10, LIDC-IDRI, and breast cancer dataset against state-of-the-art decision explanation methods DeepSHAP, DeepTaylor, and LRP (Fig. 15). In the semi-quantitative metrics introduced for the user study, namely intuitive validity ("reasonable at first glance"), semantic meaningfulness ("it makes sense"), and image quality ("visualisation looks good"), the method was superior to the benchmark methods, especially when the data were scarce. A patent was filed for this concept **(23)**.

Publications **(20-21)** present oncological imaging biomarker research with statistics and TML, where our main contribution was consultation on analysis. From experience with clinical imaging studies, we filed a patent on how to analyse such studies with statistics and AI in a fully automated manner **(22)**. This involves a large number of automated tests on the data, including heuristics from statistics and ML, and an automated generation of documentation for publication. After testing, the appropriate methods for study evaluation are selected automatically. The software described in the concept is also intended to help with study design and to propose tests to show the added value of novel technologies, such as photon-counting CT. An early subset of those automation ideas was realised by Wels et al. in a so-called radiomics prototype **(19)** which has already been used by university hospitals in a large number of clinical studies and publications.

In publication **(20)**, Woźnicki et al. combined radiomics with radiological PI-RADs criteria and clinical parameters for automated diagnosis of prostate cancer on MRI images. A combined TML model of radiomics, PI-RADs, and clinical parameters was thereby superior to PI-RADs in classifying benign vs. malignant prostate cancer and in classifying clinically significant vs. clinically insignificant prostate cancer.

Similarly, in publication **(21)** by Enke et al., radiomics applied for the spleen was used to classify malignant lymphoma as well as its subtypes by CT. Those subtypes were Hodgkin lymphoma, diffuse large B-cell lymphoma, mantle-cell lymphoma, and follicular lymphoma. It was shown that radiomics of the spleen could achieve a ROC-AUC of 0.86 for the classification of malignant lymphoma and ROC-AUCs in the range of 0.65-0.75 for the subtypes.

2.3. Image Analytics for Myo-Imaging Biomarker Research
2.3.1. Myoradiomics: Macroscopic Assessment of the Muscle-Lipid System
2.3.1.1. Building a Semi-Automated Segmentation Tool for Reproducible Muscle-Lipid Assessment of Thigh and Hip

(25) Mühlberg, A., Museyko, O., Laredo, J. D., & Engelke, K. (2017). <u>A reproducible semi-automatic method to quantify the muscle-lipid distribution in clinical 3D CT images of the thigh.</u> *PloS One*, *12*(4), e0175174.

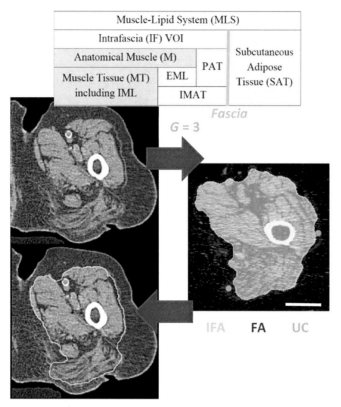

Figure 16. Compartments of the muscle-lipid system (MLS, top) and user interaction needed for semi-automated segmentation (bottom). A first coarse user estimate *G* about the status of the muscle bulk is required. The user is then shown the resulting intrafascia approximation (IFA) and the fascia approximation (FA) segmentations. Setting a few seeds by user clicks (UC) is sufficient to yield a fascia segmentation also for cases where edema and connective tissue are hard to distinguish from fascia or *moth-eaten* and *washed-out* muscle. PAT: Perimuscular AT. Bar: 10cm. Modified from Mühlberg et al.[56].

This section presents a method to process CT images of the thigh to obtain segmentations for musculoskeletal research with high reproducibility. At the time this research was conducted, novel algorithms primarily attempted to use statistical shape models or similar variants to segment individual anatomical muscles on MRI images[107-109]. Since soft tissue contrast with CT is significantly lower than with MRI, and even for MRI images, segmentation results in the published literature were not perfect, we decided against the segmentation of individual anatomical muscles and instead focussed on obtaining fast and reproducible segmentations of the muscle-lipid system (MLS, Fig. 16, top) (cf. section **1.3.1**). Also, especially before the breakthrough of the U-Net[110], most automated segmentation methods still required manual editing of the results to correct for errors.

Our idea was to allow minimal user interaction in the segmentation process, making our method semi-automated. We decided to engineer a completely deterministic method and to employ only classical image processing since the morphological variation of structures appeared too large and the sample size too small for a data-driven approach based on statistics, TML, or DL.

In a first step, the *Fascia lata* is segmented by a combination of prior information and user input. This is crucial, since underneath the fascia (intrafascia, IF), most of the relevant information regarding the mixing of lipids and muscle can be extracted.

The user, e.g., a radiologist, grades the muscle bulk by clinical criteria *moth-eaten* and *washed-out*[111] by a combined score G and within the range 1 (healthy) to 3 (pathological). This assessment is combined with a contrast metric C constructed by the CT values of the water insert from an in-scan phantom and a first coarse segmentation of adipose tissue. The metric C converts the clinical criterion *washed-out* to patient-specific CT value thresholds. This is a simple variant of an internal calibration (cf. section **2.2.1**). Also, the criterion *moth-eaten* is integrated by adapting criteria for connectedness of the muscle bulk: a 3D-region growing is applied starting from the bone surface to grow along the voxels, in whose 26-neighborhood at least (85-5 G)% of voxels have a contrast of at least C (0.75-0.15 G). By that, an initial segmentation of the muscle bulk is obtained based on the user assessment G. An anatomical coordinate system of the bone is used to obtain comparable ROIs of the thigh for each patient. Within these ROIs, the muscle bulk decomposes into connected components. Based on its size and the location relative to the bones' centre of mass, the leg of interest is selected.

For the respective leg, in a region around the muscle bulk, two segmentations are created based on the contrast C, the intrafascia approximation (IFA) and the fascia approximation (FA) (Fig. 16, bottom):

- IFA is created by repeated application of morphological operations on high contrast voxels according to C, making the segmentation compact.
- FA is created by a local-adaptive 3D region growing starting from IFA. This procedure uses contrast C and - by its locally adaptive nature - captures voxels that show higher contrast to their local neighborhood, which is a suitable prior for the detection of the thin *Fascia lata*.

If the patient is young and healthy, the outline of IFA already represents the fascia. However, for very *moth-eaten* and *washed-out* muscle and IMAT deposits between muscle and fascia, seed points can be set by user clicks (UC). Those are input to a ray reflection model[112], in which the 3D region grows along voxels surrounded by IFA and FA (Fig, 16, bottom). This criterion is tested by "emitting rays" from each voxel and fulfilled if >65% of the rays are reflected by IFA and FA. The resulting segmentation combined with IFA and FA now represents IF. The surface of IF is smoothed for the final result. Additional segmentations were created in IF. These are often directly coupled to MLS features and will, therefore, be addressed in the next section **2.3.1.2**, where also the features are introduced.

The method was evaluated for 25 patients with *washed-out* and *moth-eaten* muscle by extracting a set of features, such as volume or density, from the resulting segmentations. Inter- and intra-operator reanalysis precision for the features was assessed by application of the tool by three different users (inter-) and a three-time application (intra-) by one user. Also, the influence of calibration errors due to incorrect measurement of water as well as increased image noise was evaluated.

The results show:

- With only minor and fast user interaction (coarse grading, ca. 1-3 UC) and an internal calibration, the method provided excellent reproducibility with reanalysis errors < 1%.
- Performing a segmentation took about 12 minutes on a computer with i5 processor 5GHz, 4GB RAM, for patients with *washed-out* and *moth-eaten* muscle structures, which is an acceptable processing time.
- The method was very robust against an increase in noise level by 100%.

2.3.1.2.   Association of the Spatial Muscle-Lipid Distribution with Bone Health and Muscle Function

(26) Mühlberg, A., Museyko, O., Bousson, V., Pottecher, P., Laredo, J. D., & Engelke, K. (2019). <u>Three-dimensional distribution of muscle and adipose tissue of the thigh at CT: association with acute hip fracture</u>. *Radiology, 290*(2), pp. 426-434.

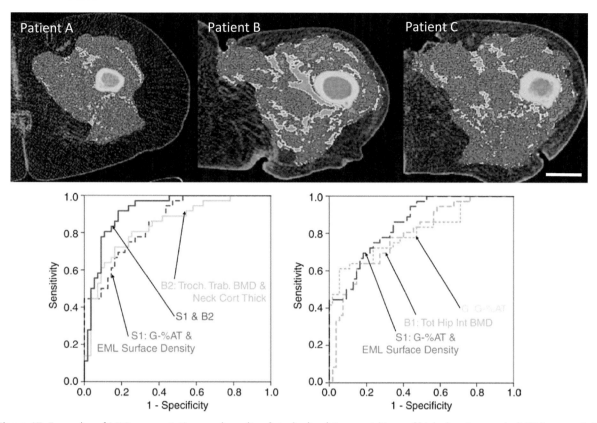

Figure 17. Examples of MLS segmentations and results of study. (top) Segmentations of high-density muscle (HDM, magenta), intermuscular adipose tissue (IMAT, yellow) and bone (green) within the fascia of the thigh for three elderly patients without (A, B) and with (C) acute fracture of the contralateral leg. (bottom) Models for fracture discrimination: (left) ROC of the best MLS model (blue), the best bone model (green) and the best combined model (magenta), and (right) comparison of the best MLS model with single adipose tissue (yellow) and bone (green) features. Bar: 10cm. Modified from Mühlberg et al.[113].

There are large numbers of CT images in hospitals initially acquired to yield more detailed information about bone health. Those images also contain complete information about the MLS, which remains unused. The idea of senior author Engelke was to use the volume of SAT for fracture discrimination.

Inspired by the emerging fields of radiomics and precision medicine for medical imaging, we expanded the project to extract extensive information about the MLS. As explained in more detail within section **1.3.1**, studies exist that associate individual components of the MLS with pathologies or muscle function, e.g., IMAT volume with endocrinological pathologies or muscle density with muscle strength. Also, there were already first studies associating bone health with muscle density and SAT volume. Our idea was to interpret these different studies as indications of one major driver: the harmful effect of muscle-lipid mixing. We quantify this driver continuously by suitable features, especially underneath the fascia. We refer to this type of radiomics for the MLS as myoradiomics.

We built upon the work presented in section **2.3.1.1.** and integrated features in the pipeline from different research branches to quantify the MLS more precisely. In addition to quantities with anatomical correspondence, like muscle tissue or (anatomical) muscle, which were segmented based on simple heuristics like connectedness and contrast, a new abstract segmentation was used: Underneath the fascia, thresholds were set for different muscle density bins based on the mean CT value of a water insert from an in-scan phantom and the mean CT value and standard deviation of SAT. By those thresholds, six bins were generated with the two extremes high-density muscle (HDM, 100% muscle; based on attenuation of the *Psoas* muscle of healthy young athletes) and IMAT (0% muscle) to enable an abstract and more detailed spatial 3D description of muscle density (Fig. 17, top and Fig. 19). EML was defined as IMAT within muscles.

Myoradiomics features were engineered that are inspired by the clinical criteria *moth-eaten* and *washed-out* to quantify the morphology underneath the fascia. For example, we employed mathematical topology and granulometry to quantify the porosity of HDM, EML, or muscle tissue segmentations. We also computed the local inhomogeneity, surface density and sphericity of those segmentations. For larger contiguous segmentations, their 3D texture was analyzed to quantify the muscle-lipid mixing, e.g., for IF, muscle, muscle tissue. For this, we extracted, e.g., fractal parameters like fractal dimension or lacunarity, but also features quantifying orientation and anisotropy like those based on a grey-value structure tensor (GST). For volume and muscle density measurement, different variants were implemented: besides the straightforward computation of density/volume of muscle or adipose tissue, in principle simply by intensity averaging/voxel counting within a segmentation, also a Gaussian mixture model (GMM) fit with Levenberg-Marquardt optimisation was applied on the CT value distribution of MLS or IF to estimate densities and volumes.

We applied the segmentation tool and the new myoradiomics features to a clinical cohort of 92 women, 40 of whom suffered an acute bone fracture. The MLS of the leg contralateral to the fracture was analysed. The objective was to use myoradiomics features for classification of fracture and test their added clinical value.

Since the sample size of 40 patients with fracture was not deemed suitable for a TML analysis, and it would have made the comparison with previous studies on bone fracture difficult, the study was analysed with statistics, making it a statistical discrimination task. For comparison with the state-of-the-art, known imaging biomarker models of bone were analysed for the same data.

By univariate statistics, a number of significant myoradiomics features were identified for this discrimination task (Table 1). Those features indicate a stronger mixing of muscle and lipids in the fracture group. Even after adjustment for age, height, and weight, muscle density was lower, muscle tissue was more porous (topology, granulometry), mixing of voxels with different densities within the muscle was stronger (variogram slope), and voxels with similar density were less oriented along the bone (GST). Additionally, high-density muscle was less oriented and more disintegrated (undirectedness, topology), and extramyocellular lipids were more finely dispersed and less oriented (inhomogeneity, surface density, undirectedness). Finally, the percentage of adipose tissue was lower in the fracture group. Interestingly, the relative volume of adipose tissue determined by the GMM-fit was significantly lower in the fracture group, but not the relative volume of SAT that is obtained by simple voxel counting.

The significant features were fed into a statistical pipeline consisting of feature elimination based on redundancy and a best subset selection according to the Bayesian Information Criterion. The five best sets were used independently with a logistic regression to yield multivariable MLS models fracture discrimination. For each model, odds ratios (ORs) for each feature and the Hosmer-Lemeshow (HL) test were employed to inspect overfitting and model robustness, respectively.

Volume of Interest and Muscle and Adipose Tissue Descriptor	Group without Fracture (n = 55)*	Group with Fracture (n = 40)*	Mean Difference[†]	Unadjusted P Value[‡]
Global percentage of adipose tissue	59.4 ± 4.7	54.3 ± 6.3	−5.1 (−8.6)	<.001 (<.001)
Intrafascial				
Muscle peak CT value (HU)	42.2 ± 7.7	37.3 ± 5.7	−4.9 (−11.6)	.001 (.03)
Standard deviation of distribution of muscle CT values (HU)	51.1 ± 8.6	52.6 ± 11.0	1.0 (1.7)	.46 (.008)
Muscle				
Variogram slope (gv)	2.2 ± 0.3	1.9 ± 0.3	−0.2 (−10.2)	<.001 (.05)
Gray-value structure tensor angle Z (radians)	1.5 ± 0.1	1.5 ± 0.1	0.02 (0.9)	.41 (.01)
Muscle tissue				
Euler density (betti/cm³)	9.6 ± 5.8	11.4 ± 7.7	1.8 (18.6)	.25 (.23)
Mean grain size (mm³)	5.0 ± 1.4	4.3 ± 1.0	−0.69 (−13.7)	.005 (.04)
High-density muscle				
Local undirectedness (gv)	79.6 ± 1.2	79.9 ± 1.1	0.33 (0.41)	.36 (.01)
Euler density (betti/cm³)	−86 ± 49	−113 ± 41	−29 (−34)	.006 (.04)
Extramyocellular lipids				
Local undirectedness (gv)	78.9 ± 0.9	79.3 ± 1.1	0.43 (0.5)	.05 (.002)
Local inhomogeneity (gv)	0.019 ± 0.07	0.023 ± 0.013	0.004 (19)	.15 (.008)
Surface density (mm)	5.0 ± 0.5	5.3 ± 0.6	0.3 (5.9)	.005 (.02)

Note.—Analysis of covariance was performed of the 12 muscle and adipose tissue descriptors that differed significantly between patients with fracture and control participants. gv = Gray value.

* Data are means ± standard deviations.

[†] Data in parentheses are percentage mean differences.

[‡] Data in parentheses are P values adjusted for age, height, and weight.

Table 1. Significant myoradiomics features in univariate statistical analysis after adjustment on epidemiologic variables age, weight, and height. Taken from Mühlberg et al.[113].

The results show (Fig. 17, bottom):

- The percentage of adipose tissue derived from a GMM and the surface density of EML were the selected features of the best MLS model. Both features were significant in the model, the confidence intervals of the ORs were fair, and the HL test indicated robustness of the model.

- Four of the five best MLS models consisted of EML surface density and GMM-based percentage of adipose tissue. The surface density feature could indicate a potentially harmful mechanism for muscle tissue when it shares a larger surface area with EML relative to the amount of EML, while the lower proportion of adipose tissue reflects frailty.

- The best MLS model was on par, ROC-AUC = 0.85 [0.78, 0.93], with the best state-of-the-art bone model, 0.84 [0.75, 0.92], for fracture discrimination.

- A combination of both models achieved a very strong fracture discrimination of 0.92 [0.86, 0.97]. Odds ratios of all features except for neck cortical thickness of the bone model were significant in the combined model.

In addition to these results, an application of myoradiomics to assess muscle function was presented at *the International Conference on Frailty & Sarcopenia Research 2015* in Boston. In this presentation, myoradiomics features were identified that showed a higher correlation with muscle quality, i.e., active force per muscle mass, than muscle density.

Our current macroscopic muscle research uses myoradiomics with statistics and NN multi-task learning to identify shared patterns of bone health and muscle function (Fig. 18). For this purpose, statistics but also decision explanation methods are utilised to rank the most important features. A manuscript draft for this is in preparation.

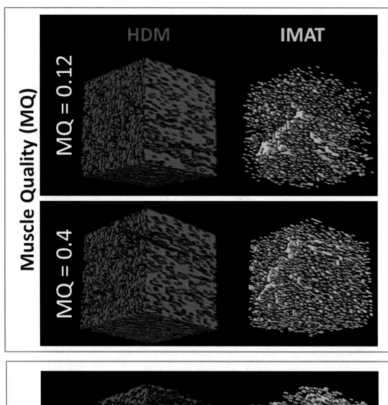

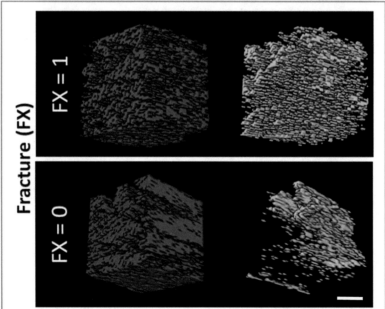

Figure 18. Comparison of structural patterns of the MLS that are associated with muscle quality (top), i.e., active force per muscle mass, and bone health (bottom). HDM and IMAT distribution are shown in a cubic *virtual biopsy*. FX = 1: Fracture. Bar: 5mm.

2.3.2. SEMPAI: A Self-Enhancing Multi-Photon Artificial Intelligence for the Identification of Microscopic Muscle Structure-Function Relationships

(27) Mühlberg A, Ritter P, Langer S, Goossens C, Nuebler S, Schneidereit D, et al. (2023). <u>SEMPAI: a Self-Enhancing Multi-Photon Artificial Intelligence for Prior-Informed Assessment of Muscle Function and Pathology.</u> *Advanced Science 10*, 2206319.

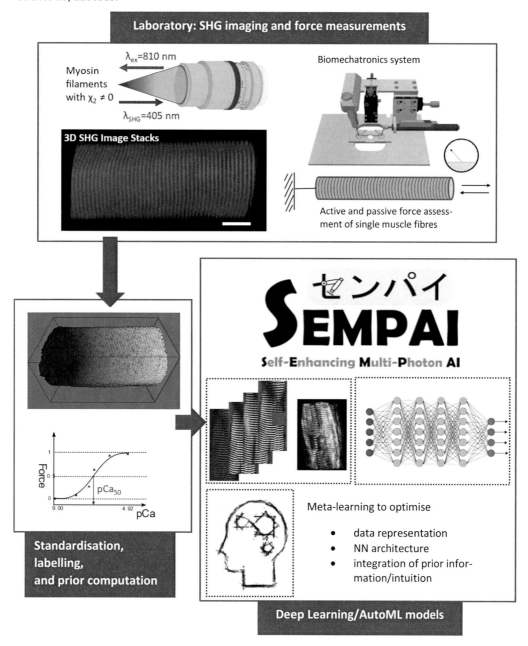

Figure 19. Processing pipeline of SEMPAI for single muscle fibres. The single fibres are imaged by SHG (top, left) and their function is assessed by a biomechatronics system (top, right). The resulting images are standardised and labels and priors are extracted from the annotated data (bottom, left). Standardised images, labels, and priors are then fed into the Self-Enhancing Multi-Photon Artificial Intelligence (SEMPAI) where DaRe, NN architecture, and prior integration are optimised by meta-learning for the prediction of multiple labels (bottom, right). Bar: 25µm. Modified from Mühlberg et al.[114].

This section presents a method to learn microscopic single muscle fibre structure-property relationships for MPM-based laboratory research. When conceptualising such an AI, requirements (Rq.) can be defined as a lesson of the results in the last sections and general considerations. First, the required model complexity is not known, and whether statistics, TML, or DL should be employed. Thus, an agnostic approach is required (Rq. 1). We have learned about the importance of DaRe (section **2.1.1**), and the usefulness of prior information (sections **2.2**, **2.3**) for prediction tasks. The appropriate DaRe for a task is most often unknown and, as seen, should not be chosen arbitrarily (Rq. 2). Although DL is also called representation learning, the choice of the DaRe that is entered in an NN is in practice still important. Another aspect is that the objective of fundamental medical research is often not solely to improve a clinical diagnosis but also to gain more general knowledge. Accordingly, an AI should provide even more information than in macroscopic imaging in radiology (Rq. 3). In addition, the researcher often already has intuition or knowledge about a problem, so it might be helpful to at least be able to offer this prior knowledge to the AI without limiting it (Rq. 4). Concluding, it is crucial to construct an AI that directly interacts with the researcher. The last requirement is addressing an optimal utilisation of data: specifically in laboratory experiments, there is often only small sample size and parts of information might be missing. Therefore, it is important to enable the analysis of smaller experiments and heterogeneous, sparse laboratory data (Rq. 5).

Based on the defined requirements, our main idea was to optimise the DaRe together with the NN architecture and prior integration. Protocolling the optimisation of DaRe provides the researcher with important insights about the origin of biological information (cf. Rq. 2 & Rq. 3). The priors that are offered by the researcher in form of handcrafted features can be integrated by the AI, but the AI can also reject their use. Accordingly, this can be utilised like a hypothesis test and allows the biomedical scientist to actively participate in research (cf. Rq. 4). The NN architecture and its hyperparameters, or even more general the ML approach, adapt to DaRe and prior integration, thereby adapting the complexity of the method (cf. Rq. 1) to the prediction task. As our second basic idea, the AI is conceptualised for joint learning, i.e., new experiments are added to the data pool of older ones. By this, it is expected that less data are needed for prediction tasks of a new experiment while the older experiments benefit from the new data as well (cf. Rq. 5). This approach is also the basic principle of transfer and multi-task learning. Based on the expected enhancement of the AI over the course of time by adding more data and due to the optimisation, we, therefore, term it Self-Enhancing Multi-Photon AI (SEMPAI). A scheme of the whole processing pipeline is shown in Fig. 19.

For a practical implementation, we first had to collect, curate, and standardise experimental data that had been acquired over a decade at the Institute of Medical Biotechnology (MBT). Since the data were spread across the file system of the server and varied in format and quality, a lot of effort had to be invested here: for data engineering, automated data collection and standardisation pipelines were engineered for each experiment, and standardisation was protocolled using an automatically generated SEMPAI *Labbook* for rapid communication among researchers and discussion of potential revision or exclusion of data. The standardisation aims to minimise the technical variation of the images between experiments (cf. section **2.2.1**) and to orient and position them as identically as possible for DL. To minimise technical variation, different denoising algorithms were explored but used only moderately to conserve fine-grained biological information. Subsequently, after resampling to a uniform voxel size and contrast enhancement, the images were registered to a muscle fibre with particularly advantageous properties using a rigid multi-scale registration. After normalisation of the images, an average image was determined by superposition. A bounding box was created for the threshold-based segmentation of this average image. All images were then cropped to the generated bounding box to minimise the dimensionality of the input data (*curse of dimensionality*, cf. section **1.2**) (Fig. 20).

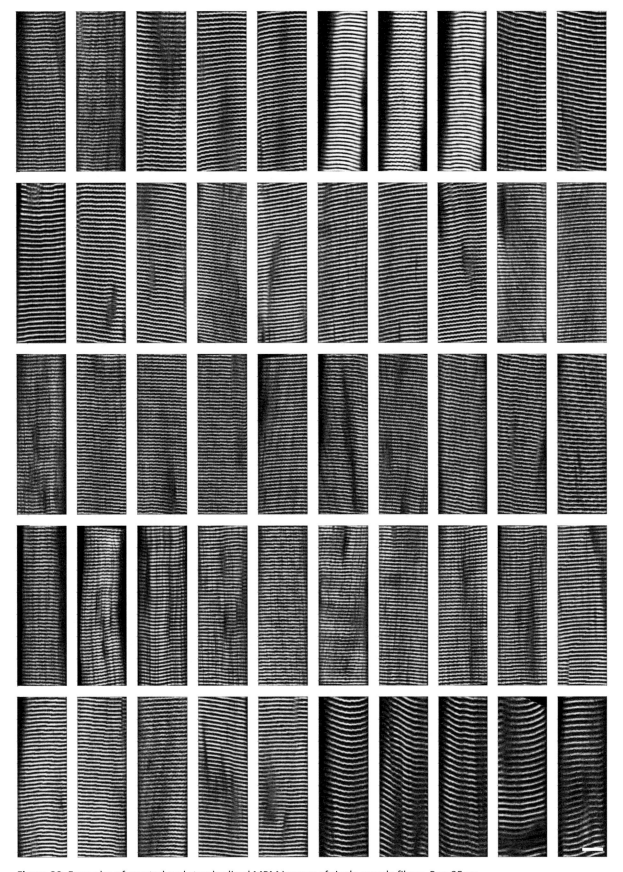

Figure 20. Examples of curated and standardised MPM images of single muscle fibres. Bar: 25µm.

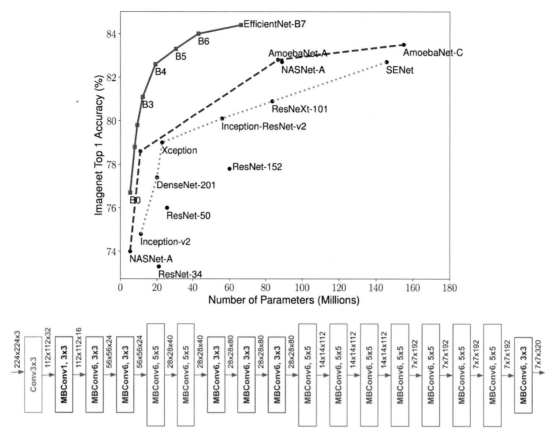

Figure 21. (top) Benchmarking of EfficientNet variants B0-B7 against some of the most popular NN architectures for ImageNet database as a function of NN parameters, i.e., degrees of freedom. (bottom) Architecture of EfficientNet-B0. Modified from Tan et al.[115].

Since the techniques to quantify the cross-sectional area (CSA) of the muscle fibre varied between the original experiments, a novel method was engineered, which re-orientates the fibre vertically and combines three different algorithms for CSA computation. By this approach, outliers in the CSA determination can be identified and excluded, and results are more robust by averaging of the results.

Data with the required quality, as identified by *Labbook* inspection, were fed into SEMPAI together with the priors, i.e., the best current imaging biomarkers in the single muscle fibre research, and the labels, i.e., the pathologies and functional properties to be predicted. Examples of standardised fibres are shown in Fig. 20. In total, we managed to curate and standardise 1,298 datasets. To the best of our knowledge, this makes it the largest data collection of single muscle fibres for which AI has been used to date.

SEMPAI employs meta-learning to simultaneously optimise DaRe, NN architecture, and prior integration. For this purpose, it utilises the multi-objective NSGA-II[116] algorithm and a simple early stopping criterion for pruning. Although the very high computational cost of meta-learning has to be reduced, pruning must not be too strict, otherwise trials are not comparable, and it is impossible to draw conclusions about optimal meta-configurations. Regarding DaRe of the samples, SEMPAI can choose between 2.5D, i.e., three representative slices of selectable distance to each other, or 3D and can decide to downsample the image resolution, to apply a contrast enhancement algorithm, to erase parts of the image (random erasing), and to augment the images. For the NN architecture, based on our yet small sample size for DL and the *curse of dimensionality*, we aimed to use a set of variants that have been previously shown to yield good predictive performance with rather few parameters. We, therefore, integrated EfficientNet[115] which has the additional advantage of having scaled variants

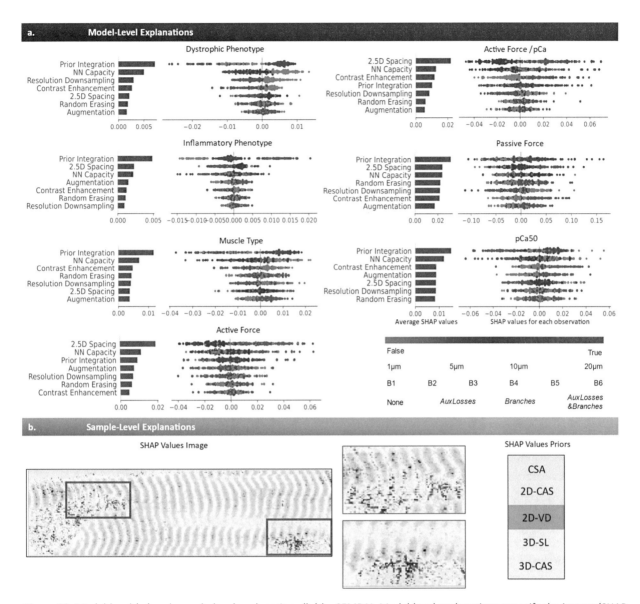

Figure 22. Model-level (a.) and sample-level explanations (b.) by SEMPAI. Model-level explanations quantify the impact (SHAP values) of different configurations (colour coded, legend on bottom right of a.) on predictive performance. This shows, for instance, that a strong integration of priors (as auxiliary losses and branches) had positive impact on the prediction of inflammatory phenotype. Sample-level explanations simultaneously visualise the impact (SHAP values) of image voxels (left) and priors (right) of each sample individually for the prediction of labels. In this example, the relevant voxels and priors for the prediction of dystrophic phenotype (cf. **section 1.3.1**) are visualised by colour overlay. This shows that twisted regions in the fibre and the prior 2D-VD are important to correctly classify this sample as dystrophic. Priors: Cross-Sectional Area (CSA), 2D/3D-Cosinus Angle Sum (CAS), 2D-Vernier Density (VD), 2D-Sarcomere Length (SL). Taken from Mühlberg et al.[114].

with different capacity/complexity (Fig. 21). SEMPAI can select one of the architectures B1-B6 of EfficientNet. In addition to simultaneous tuning of NN hyperparameters like batch size or learning rate, NN techniques such as importance sampling (cf. section **2.2.3**) or gradient clipping are available for SEMPAI to use. Finally, SEMPAI can decide to integrate prior information but can also reject its use. Several variants of prior integration were implemented: SEMPAI can use the priors as auxiliary tasks, add them as branches to the fully connected layer, or both. In addition, SEMPAI has the option to use AutoML with priors only. For this, TPOT-SH[117] is integrated, which contains a variety of feature selection methods and TML classifiers/regressors, some of which are also common models in statistics. By this, SEMPAI can select the optimal method from the whole continuum from statistics over TML to DL with and without priors, making it an agnostic approach w.r.t. learning method selection.

SEMPAI learns all tasks simultaneously. By this joint learning, the use of shared patterns is enforced to be able to make generalising predictions also for tasks with fewer samples[118]. The use of shared patterns can also help to reduce the impact of technical variation as those patterns are less probable to be technical in nature. The losses for each task are weighted by uncertainty weighting[119]. Data heterogeneity is addressed by masking out losses of labels when this information is missing for the respective sample while still using all available labels.

The researcher obtains feedback on whether the provided priors were important and in which DaRe the relevant information can be found. The meta-parameters of the models that were protocolled during meta-learning are associated with the predictive performance provided by a TML model. The learned association is explained by SHAP Tree Explainer[120] (model-level explanations, Fig. 22, top). In addition, DL models are analysed with DeepSHAP[121]. For the DL models that integrated priors as branches, the importance of priors is shown simultaneously with the importance of image voxels/pixels (sample-level explanations, Fig. 22, bottom) to yield deeper insights into their relative importance.

The results show:

- SEMPAI provided better predictive performance than the state-of-the-art imaging biomarkers in six of seven tasks, including the prediction of active and passive force and dystrophic phenotype.
- Integration of priors into DL yielded the best results, better than DL without priors or TML with priors only. This can either be interpreted as a DL-based-refinement of prior information by additional image information or a regularisation of DL by priors.
- Joint learning in SEMPAI allowed to find DL models with good predictive performance even for tasks with very small sample size.
- The researcher obtains relevant knowledge about the origin of biological information. SEMPAI provided systematic feedback, e.g., that most of the diagnostic information for the inflammatory phenotype is already provided by a prior, that contrast enhancement in dystrophic fibres destroys relevant information, or that relevant information about muscle force can be obtained in the periphery of the muscle fibre.

As prospect, it is planned to apply SEMPAI for MPM colon data. For this, appropriate priors have to be formulated. Although we have already spent much effort to speedup SEMPAI's meta-learning by a variety of optimisation steps, including automatic mixed precision, gradient accumulation, or augmentation on the GPU, we aim to decrease SEMPAI's computational cost by using further techniques from NAS research such as weight inheritance or performance estimation strategies.

In the medium term, we plan to implement SEMPAI 2.0 for muscle fibres. This will employ known operator learning by integrating a local Fourier transform layer into the DL architectures of SEMPAI. This seems reasonable since most priors are based on spatial frequency in order to effectively quantify the striated single muscle fibre structure. By introduction of this layer, we are expecting a further *self-enhancement* for the identification of imaging biomarkers in myology.

3. Conclusions and Outlook

In the present book, we have developed **image processing and prior-informed learning techniques** to identify **novel macroscopic and microscopic imaging biomarkers for pneumonology, oncology, and myology**.

While it has generally been shown that DL is the method of choice when big data are available, the *curse of dimensionality* often makes the training of such models hard in practice when there is a small sample size. Especially in medical research, there are often not enough samples to identify generalising biomarkers with DL without any **domain knowledge**. Indirectly, the **integration of priors** can be considered an information transfer from studies with more data where those priors were initially identified, thereby helping to avoid **overfitting**. For example, in section **2.3.2**, integrated priors were already relevant in studies on single muscle fibre structure-function relationships. There is also a research trend to integrate priors as layers or as auxiliary losses in DL with partly excellent success[38,122]. The branch of scientific machine learning is an emerging research field in which hybrid approaches of classical scientific modelling and AI are proposed. An example of this branch are physics-informed neural networks, where input and output of the NN have to satisfy physics differential equations, thereby using prior information to regularise the NNs[123]. As another advantage, the utilisation of priors can lead to **more interpretable models**. As Rudin et al. stated in an article in Nature Artificial Intelligence, it might be beneficial to focus on making models interpretable from the beginning instead of explaining them afterwards[124]. The results in this work show that combining priors with TML and especially DL can yield models with superior predictive performance. In section **2.3.2**, the best models for the prediction of single muscle fibre properties integrated priors in DL. Those models were superior to naïve DL without priors, TML models based solely on priors, and state-of-the-art imaging biomarkers. Similarly, in section **2.2.1**, physical priors, in the form of qualification metrics for surrogates, helped to calibrate known texture biomarkers in TML against technical variation, and this approach was superior to naïve radiomics. In section **2.1.1**, a TML model of prior-knowledge inspired features and optimised DaRe was superior to an advanced 3D DL algorithm - which was even pre-trained on a large cohort - for COPD phenotyping.

Another implication of this work, which at first glance seems to contradict the last paragraph but will be clarified later, is that one should **construct the AI system agnostically**. Based on our results, it is suboptimal to make an early decision regarding the choice of methods (statistics vs. TML vs. DL), and the DaRe (which is often part of the study design), as shown in sections **2.1.1** and **2.3.2**. **The impact of the DaRe on the predictive performance can be larger than the learning algorithm** as seen in section **2.1.1**. Therefore, especially DaRe and study design selection should be part of the optimisation. **Meta-learning** or AutoML are interesting variants to achieve this. Their application for optimisation of NN architecture design in the form of NAS has already shown its benefits. It should be pointed out that optimising the study design for TML and DL might be different to an optimisation for radiological reading, because findings from radiology regarding DaRe may not necessarily be transferable to learning-based methods as shown in section **2.1.1**. Accordingly, prior knowledge about a good study design or DaRe specifically for AI applications first has to be generated.

Regarding the consideration of using models that use **prior knowledge versus models as a result of meta-learning**, both of which have shown their benefit in this work, **different perspectives exist in computer science**. For instance, Sutton[125] states that in the near future, all AI model construction should be based on meta-learning while constraining models by prior knowledge may result in missing optimal solutions as observed in chess: *"We want AI agents that can discover like we can, not which contain what* *we have discovered. Building in our discoveries only makes it harder to see how the discovering process can be done.*" A similar argument was raised by Hosny et al. when a DL model for radiotherapy integrated the Aerts signature as a prior[126]. Maier et al.[40], on the other hand, argue that meta-learning often fails to effectively identify inductive bias and that the integration of prior knowledge about the target function is the only way to minimise bias and variance simultaneously[39,40]. They state: *"As a result, particularly successful machine learning models encode knowledge about their application in order to outperform other approaches quite frequently"*. In support of their claim, the method AlphaFold 2[127], which predicts the tertiary structure of a protein as a function of its primary structure, *"incorporates physical and biological knowledge about protein structure"*[127] and has probably been the largest success of AI in biomedicine, or even in general, to date.

In this work, we propose a **hybrid approach of both philosophies**. SEMPAI (section **2.3.2**) employs meta-learning across tasks and determines the level of prior integration as part of the optimisation process, from none to exclusively relying on priors. To the best of our knowledge, the level of prior integration was previously a design choice and not part of an optimisation from an agnostic starting point.

From a **biomedical perspective**, this work addresses the **expansion of precision medicine to further domains**. While DL is susceptible to overfitting, thus requiring tailored solutions and profiting from regularisation by priors, there is evidence in this work for research branches with previous **underfitting** due to insufficient information extraction. For example, our novel approach myoradiomics (section **2.3.1**) to associate the macroscopic MLS with bone health and muscle function is promising. Previously, only muscle density and volumes of different muscles and AT deposits were quantified by CT. Myoradiomics includes a semi-automatic segmentation algorithm and a prior information-based set of features to describe the spatial mixing of muscle and lipids underneath the fascia in a continuous manner. Using myoradiomics in combination with a statistical approach adapted to the sample size, we were able to identify novel relevant myoradiomics features and significantly improve fracture discrimination compared to common bone biomarker models alone. Similarly, our novel approach trajectomics (section **2.2.2**) builds upon biological knowledge that distant metastases are associated with poor survival outcome, translates this prior to organ level and combines it with TML and geometric DL. This is a step towards fine-grained personalised medicine based on macroscopic tumour growth trajectories. In addition, this approach offers the advantage over conventional texture-based DL analysis of tumours that it is not, or at least less, susceptible to technical variation.

We have worked on concepts for **DL variants of technome and trajectomics**, i.e., DeepTechnome and DeepTrajectomics, since it is expected that those prior-informed DL-based methods will perform better than TML for a large scale industrial application when big data can be acquired. Those concepts have been filed as patents **(7, 12)** and their vanilla versions have already been explored and implemented under our supervision by authors Langer **(5)** and Rist **(10)** with promising first results.

The **outlook** of the research presented in this book was already partially outlined in the context of the theme of the respective section, but it will be expanded and put more in context here.

First, we discuss the planned projects for our macroscopic biomarker research. In our DeepTechnome research, we will both explore further aspects of the granted patent **(7)**, like the integration of (noise) simulations, and investigate their application for MRI and histological images. We also compare our developments with adversarial debiasing[128] which, unlike DeepTechnome, debiases from *known technical bias*, i.e., the external labels, e.g., from the DICOM information, such as kernel or scanner model. DeepTechnome does not require those technical labels but identifies yet *unknown technical bias* from

CRs. This is important, because there is no external information for artifacts or noise, especially not specific to the relevant technical variation of the respective task. It is, however, possible that we will extend our DeepTechnome method with insights from the field of adversarial debiasing. For example, DeepTechnome could only be employed to identify the technical bias from the control regions, while adversarial debiasing is then utilised to calibrate for the identified bias. We also aim to further explore the concept presented in the filed patent **(2)**. For this, advanced NAS technologies will be implemented in combination with simultaneous DaRe optimisation. The hope is that DaRes can be identified that are tailored for COPD research. This may then subsequently lead to the identification of even better automated COPD phenotyping systems. The large data set of 981 patients still provides a very good foundation for this objective. Since this data set is also used for DeepTechnome development, an integrated solution of both methods may emerge, i.e., a method that both optimises the DaRe and internally calibrates the DL system. The large importance of pan-cancer research in oncology and our positive experience with the effective utilisation of shared patterns by SEMPAI drive our continued work to identify shared patterns in our trajectomics and myoradiomics research. This research, i.e., pan-myoradiomics and pan-trajectomics, will not only employ techniques already known from SEMPAI, like multi-task learning and decision explanation, but will also explore -omics-techniques to identify structural patterns that are associated with more than one prediction task.

For our microscopic biomarker research, we will maintain the SEMPAI framework and extend it modularly. For this purpose, a workstation with version control has been set up at MBT. The next implementation steps for SEMPAI 2.0 are already planned. High priority is given to a local Fourier transform layer, integration of more NN architectures, and optimisation techniques with lower computational cost borrowed from NAS. The large amount of effort that has gone into rather technical aspects, such as acceleration of training, a variety of data standardisation tools as well as documentation of data collection and standardisation by automated *Labbooks,* is needed for the bigger picture. The long-term vision behind SEMPAI is to construct an AI into which standardised data from new experiments can be fed without much additional effort by the experimenter and which automatically interprets these data by the knowledge from previous experiments. Meta-learning then systematically identifies similarities and differences between various diseases and experiments. Especially the optimisation of the network architecture which is often done manually by trial and error —resulting in the humorous term *graduate student descent*[40]-, will be automatically performed by SEMPAI's integrated NAS techniques and result in a significant boost in productivity. Also, using all data of different tasks simultaneously for training should pay off in the long run, especially for drawing conclusions from smaller experiments. Many more data from MBT yet have to be integrated into SEMPAI's training data. A shared, large standardised database then not only enables the yet employed joint learning with all its advantages, but many different paths can be explored. For instance, GANs (cf. section **2.2.3**) can be integrated into SEMPAI. Once a GAN has been trained on big data, e.g., of muscle fibres with various labels, it can also be used for style transfer between phenotype or to synthesise[129] new virtual muscle fibres of specific phenotypes and function. These data can then be used, for example, to augment the training data (cf. **(24)**). Finally, it is conceivable to integrate simulations, e.g., finite element modelling[130] of muscle fibres, as *in silico* priors. This would enable further approaches, such as augmentation with simulated data, utilisation of PINNs[123] to predict the outcome of biomechatronics experiments with a strong physics rationale, and *in silico* plausibility/sanity checks of experimental or AI results. At this stage, SEMPAI would represent a digital twin[131,132] of biological structures in which experiment, simulation, and AI complement, verify, and *enhance* each other.

4. References

1   U.S. Food & Drug Administration. *Computed Tomography (CT)*, <https://www.fda.gov/radiation-emitting-products/medical-x-ray-imaging/computed-tomography-ct> (2023).

2   Van Aarle, W. *et al. Computed tomography and ASTRA Toolbox training course*, <https://visielab.uantwerpen.be/computed-tomography-and-astra-toolbox-training-course> (2016).

3   Kalender, W. A. *Computed Tomography: Fundamentals, System Technology, Image Quality, Applications*. (John Wiley & Sons, 2011).

4   Radon, J. On the determination of functions from their integral values along certain manifolds. *IEEE Transactions on Medical Imaging* **5**, 170-176 (1986).

5   Brooks, R. A. & Di Chiro, G. Theory of image reconstruction in computed tomography. *Radiology* **117**, 561-572 (1975).

6   Beister, M., Kolditz, D. & Kalender, W. A. Iterative reconstruction methods in X-ray CT. *Physica Medica* **28**, 94-108 (2012).

7   Willemink, M. J. *et al.* Iterative reconstruction techniques for computed tomography Part 1: technical principles. *European Radiology* **23**, 1623-1631 (2013).

8   Schöckel, L. *et al.* Developments in X-ray contrast media and the potential impact on computed tomography. *Investigative Radiology* **55**, 592-597 (2020).

9   Momcilovic, M. & Shackelford, D. B. Imaging cancer metabolism. *Biomolecules & Therapeutics* **26**, 81 (2018).

10   Litjens, G., van Laarhoven, C. J. H. M., Prokop, M., van Geenen, E. J. M. & Hermans, J. J. Routine contrast-enhanced CT is insufficient for TNM-staging of duodenal adenocarcinoma. *Abdominal Radiology* **47**, 3436-3445 (2022).

11   Schroeder, S. *et al.* Cardiac computed tomography: indications, applications, limitations, and training requirements: report of a Writing Group deployed by the Working Group Nuclear Cardiology and Cardiac CT of the European Society of Cardiology and the European Council of Nuclear Cardiology. *European Heart Journal* **29**, 531-556 (2008).

12   Marin, D., Boll, D. T., Mileto, A. & Nelson, R. C. State of the art: dual-energy CT of the abdomen. *Radiology* **271**, 327-342 (2014).

13   Willemink, M. J., Persson, M., Pourmorteza, A., Pelc, N. J. & Fleischmann, D. Photon-counting CT: technical principles and clinical prospects. *Radiology* **289**, 293-312 (2018)

14   Clark, D. P. & Badea, C. Micro-CT of rodents: state-of-the-art and future perspectives. *Physica Medica* **30**, 619-634 (2014).

15   Matsui, T. *et al.* Non-labeling multiphoton excitation microscopy as a novel diagnostic tool for discriminating normal tissue and colorectal cancer lesions. *Scientific Reports* **7(1)**, 1-9 (2017).

16   Kreiss, L. *et al.* Label-free analysis of inflammatory tissue remodeling in murine lung tissue based on multiphoton microscopy, Raman spectroscopy and machine learning. *Journal of Biophotonics* **15**, e202200073 (2022).

17   Denk, W., Strickler, J. H. & Webb, W. W. Two-photon laser scanning fluorescence microscopy. *Science* **248**, 73-76 (1990).

18   Zipfel, W. R., Williams, R. M. & Webb, W. W. Nonlinear magic: multiphoton microscopy in the biosciences. *Nature Biotechnology* **21**, 1368-1376 (2003).

19   Dempsey, W. P., Fraser, S. E. & Pantazis, P. SHG nanoprobes: Advancing harmonic imaging in biology. *BioEssays* **34**, 351-360 (2012).

20   Both, M. *et al.* Second harmonic imaging of intrinsic signals in muscle fibers in situ. *Journal of Biomedical Optics* **9**, 882-892 (2004).

21   Schürmann, S., Weber, C., Fink, R. H. & Vogel, M. Myosin rods are a source of second harmonic generation signals in skeletal muscle. in *Multiphoton Microscopy in the Biomedical Sciences VII*, 310-315 (International Society for Optics and Photonics).

22   Nucciotti, V. *et al.* Probing myosin structural conformation in vivo by second-harmonic generation microscopy. *Proceedings of the National Academy of Sciences* **107**, 7763-7768 (2010).

23   Dilipkumar, A. *et al.* Label-free multiphoton endomicroscopy for minimally invasive in vivo imaging. *Advanced Science* **6(8)**, 1801735 (2019).

24   Encyclopedia Britannica. *Statistics*, <https://www.britannica.com/science/statistics> (2023).

25   Broemeling, L. D. An account of early statistical inference in Arab cryptology. *The American Statistician* **65**, 255-257 (2011).

26   Walker, H. M. *Studies in the History of Statistical Method* (The William & Wilkins Company, 1929).

27   Hastie, T., Tibshirani, R. & Friedman, J. H. *The elements of statistical learning: data mining, inference, and prediction*. Vol. 2 (Springer, 2009).

28   Samuel, A. L. Some studies in machine learning using the game of checkers. II—Recent progress. *IBM Journal of Research and Development* **11**, 601-617 (1967).

29   Bzdok, D., Altman, N. & Krzywinski, M. Statistics versus Machine Learning. *Nature Methods* **15**, 233-234 (2018).

30   Aerts, H. J. *et al.* Decoding tumour phenotype by noninvasive imaging using a quantitative radiomics approach. *Nature Communications* **5**, 1-9 (2014).

31   Gillies, R. J., Kinahan, P. E. & Hricak, H. Radiomics: images are more than pictures, they are data. *Radiology* **278**, 563 (2016).

32   Cybenko, G. Approximation by superpositions of a sigmoidal function. *Mathematics of Control, Signals and Systems* **2**, 303-314 (1989).

33   Ciregan, D., Meier, U. & Schmidhuber, J. Multi-column deep neural networks for image classification. in *2012 IEEE Conference on Computer Vision and Pattern Recognition,* 3642-3649 (IEEE).

34   Krizhevsky, A., Sutskever, I. & Hinton, G. E. Imagenet classification with deep convolutional neural networks. *Communications of the ACM* **60**, 84-90 (2017).

35   Sennaar, K. *How America's 5 Top Hospitals are Using Machine Learning Today*, <https://emerj.com/ai-sector-overviews/top-5-hospitals-using-machine-learning/> (2023).

36   Lapuschkin, S. *et al.* Unmasking clever hans predictors and assessing what machines really learn. *Nature Communications* **10**, 1-8 (2019).

37   Simonyan, K., Vedaldi, A. & Zisserman, A. Deep inside convolutional networks: Visualising image classification models and saliency maps. *arXiv Preprint arXiv:1312.6034* (2013).

38   Maier, A. K. *et al.* Learning with known operators reduces maximum error bounds. *Nature Machine Intelligence* **1**, 373-380 (2019).

39   Hart, P. E., Stork, D. G. & Duda, R. O. *Pattern Classification*. (Wiley, 2000).

40   Maier, A., Köstler, H., Heisig, M., Krauss, P. & Yang, S. H. Known operator learning and hybrid machine learning in medical imaging—a review of the past, the present, and the future. *Progress in Biomedical Engineering* **4(2)**, 022002.

41   Elsken, T., Metzen, J. H., & Hutter, F. Neural architecture search: A survey. *The Journal of Machine Learning Research* **20(1)**, 1997-2017 (2019).

42   Bellman, R. Dynamic programming. *Science* **153**, 34-37 (1966).

43   Theodoridis, S. & Koutroumbas, K. *Pattern Recognition*. (Elsevier, 2006).

44   Poggio, T., Mhaskar, H., Rosasco, L., Miranda, B. & Liao, Q. Why and when can deep-but not shallow-networks avoid the curse of dimensionality: a review. *International Journal of Automation and Computing* **14**, 503-519 (2017).

45   Ng, A. Sparse autoencoder. *CS294A Lecture Notes* **72**, 1-19 (2011).

46   Saha, S. *A Comprehensive Guide to Convolutional Neural Networks—the ELI5 way*, <https://towardsdatascience.com/a-comprehensive-guide-to-convolutional-neural-networks-the-eli5-way-3bd2b1164a53> (2022).

47   Mukund, K. & Subramaniam, S. Skeletal muscle: A review of molecular structure and function, in health and disease. *Wiley Interdisciplinary Reviews: Systems Biology and Medicine* **12**, e1462 (2020).

48   Schmidt-Nielsen, K. *Physiologie der Tiere*. (Spektrum Akad. Verlag, 1999).

49   Huxley, A. & Niedergerke, R. Interference microscopy of living muscle fibres. *Nature* **173**, 13 (1954).

50   Huxley, A. F. Muscle structure and theories of contraction. *Progress in Biophysics and Biophysical Chemistry* **7**, 256-319 (1957).

51   Kandel, E. R. *et al. Principles of Neural Science*. (McGraw-Hill New York, 2000).

52   Hwang, P. M. & Sykes, B. D. Targeting the sarcomere to correct muscle function. *Nature Reviews Drug Discovery* **14**, 313-328 (2015).

53   Katirji, B. Clinical assessment in neuromuscular disorders. *Neuromuscular Disorders in Clinical Practice*, 3-20 (2014).

54   Duan, D., Goemans, N., Takeda, S. i., Mercuri, E. & Aartsma-Rus, A. Duchenne muscular dystrophy. *Nature Reviews Disease Primers* **7**, 1-19 (2021).

55   Emery, A. E. Population frequencies of inherited neuromuscular diseases—a world survey. *Neuromuscular Disorders* **1**, 19-29 (1991).

56   Mühlberg, A., Museyko, O., Laredo, J.-D. & Engelke, K. A reproducible semi-automatic method to quantify the muscle-lipid distribution in clinical 3D CT images of the thigh. *PloS One* **12**, e0175174 (2017).

57   Frost, H. M. Bone "mass" and the "mechanostat": a proposal. *The Anatomical Record* **219**, 1-9 (1987).

58   Waters, D. L. Intermuscular adipose tissue: a brief review of etiology, association with physical function and weight loss in older adults. *Annals of Geriatric Medicine and Research* **23**, 3 (2019).

59   Goodpaster, B. H., Bergman, B. C., Brennan, A. M. & Sparks, L. M. Intermuscular adipose tissue in metabolic disease. *Nature Reviews Endocrinology* **19**, 285–298 (2023).

60   Biltz, N. K. *et al.* Infiltration of intramuscular adipose tissue impairs skeletal muscle contraction. *The Journal of Physiology* **598**, 2669-2683 (2020).

61   Coen, P. M. & Goodpaster, B. H. Role of intramyocelluar lipids in human health. *Trends in Endocrinology & Metabolism* **23**, 391-398 (2012).

62   Parkkari, J. *et al.* Energy-shunting external hip protector attenuates the peak femoral impact force below the theoretical fracture threshold: An in vitro biomechanical study under falling conditions of thel elderly. *Journal of Bone and Mineral Research* **10**, 1437-1442 (1995).

63   Goodpaster, B. H., Kelley, D. E., Thaete, F. L., He, J. & Ross, R. Skeletal muscle attenuation determined by computed tomography is associated with skeletal muscle lipid content. *Journal of Applied Physiology* **89**, 104-110 (2000).

64   Larson-Meyer, D. E. *et al.* Muscle-associated triglyceride measured by computed tomography and magnetic resonance spectroscopy. *Obesity* **14**, 73-87 (2006).

65   Daguet, E. *et al.* Fat content of hip muscles: an anteroposterior gradient. *The Journal of Joint & Bone Surgery* **93**, 1897-1905 (2011).

66   Cesari, M. *et al.* Biomarkers of sarcopenia in clinical trials—recommendations from the International Working Group on Sarcopenia. *Journal of Cachexia, Sarcopenia and Muscle* **3**, 181-190 (2012).

67   Lang, T. *et al.* Computed tomographic measurements of thigh muscle cross-sectional area and attenuation coefficient predict hip fracture: the health, aging, and body composition study. *Journal of Bone and Mineral Research* **25**, 513-519 (2010).

68   National Institute of Cancer. *What Is Cancer?*, <https://www.cancer.gov/aboutcancer/understanding/what-is-cancer> (2021).

69   Gebauer, L. *et al.* Quantitative Imaging Biomarkers of the Whole Liver Tumor Burden Improve Survival Prediction in Metastatic Pancreatic Cancer. *Cancers* **13**, 5732 (2021).

70   Sung, H. *et al.* Global cancer statistics 2020: GLOBOCAN estimates of incidence and mortality worldwide for 36 cancers in 185 countries. *CA: a Cancer Journal for Clinicians* **71**, 209-249 (2021).

71   Eisenhauer, E. A. *et al.* New response evaluation criteria in solid tumours: revised RECIST guideline (version 1.1). *European Journal of Cancer* **45**, 228-247 (2009).

72   Edge, S. B. & Compton, C. C. The American Joint Committee on Cancer: the 7th edition of the AJCC cancer staging manual and the future of TNM. *Annals of Surgical Oncology* **17**, 1471-1474 (2010).

73   Lemeshow, S., May, S. & Hosmer Jr, D. W. *Applied Survival Analysis: Regression Modeling of Time-to-Event Data*. (John Wiley & Sons, 2011).

74   Hu, T., Kumar, Y., Ma, E. Z., Wu, Z., & Xue, H. Pan-cancer analysis of whole genomes. *Nature* **578**, 82-93 (2020).

75   Srivastava, S. & Hanash, S. Pan-cancer early detection: hype or hope? *Cancer Cell* **38**, 23-24 (2020).

76   Betts, J. G. *et al. Anatomy and Physiology*. (Rice University, 2013).

77   Letko, M., Marzi, A. & Munster, V. Functional assessment of cell entry and receptor usage for SARS-CoV-2 and other lineage B betacoronaviruses. *Nature Microbiology* **5**, 562-569 (2020).

78   Huang, C. *et al.* Clinical features of patients infected with 2019 novel coronavirus in Wuhan, China. *The Lancet* **395**, 497-506 (2020).

79   Torres-Castro, R. *et al.* Respiratory function in patients post-infection by COVID-19: a systematic review and meta-analysis. *Pulmonology* **27**, 328-337 (2021).

80   Adeloye, D. *et al.* Global, regional, and national prevalence of, and risk factors for, chronic obstructive pulmonary disease (COPD) in 2019: a systematic review and modelling analysis. *The Lancet Respiratory Medicine* **10(5)**, 447-458 (2022).

81   Mühlberg, A. *et al.* Unraveling the Interplay of Image Formation, Data Representation and Learning in CT-based COPD Phenotyping Automation: The Need for a Meta-Strategy. *Medical Physics* **48(9)**, 5179-5191 (2021).

82   Olson, R. S. & Moore, J. H. TPOT: A Tree-Based Pipeline Optimization Tool for Automating Machine Learning. *Automated Machine Learning*, 151-160 (Springer, 2019).

83   Odry, B. *et al.* Pulmonary lobe segmentation using the thin plate spline (TPS) with the help of the fissure localization areas. in *Medical Imaging 2013: Image Processing*, 244-254. (International Society for Optics and Photonics).

84   Odry, B. L., Kiraly, A. P., Novak, C. L., Naidich, D. P. & Lerallut, J.-F. Automated airway evaluation system for multi-slice computed tomography using airway lumen diameter, airway wall thickness and broncho-arterial ratio. in *Medical Imaging 2006: Physiology, Function, and Structure from Medical Images*, 243-253 (International Society for Optics and Photonics).

85   Seifert, S. *et al.* Hierarchical parsing and semantic navigation of full body CT data. in *Medical Imaging 2009: Image Processing* (International Society for Optics and Photonics).

86   Chen, H., Dou, Q., Yu, L., Qin, J. & Heng, P.-A. VoxResNet: Deep voxelwise residual networks for brain segmentation from 3D MR images. *NeuroImage* **170**, 446-455 (2018).

87   Ahmed, J. *et al.* COPD Classification in CT Images Using a 3D Convolutional Neural Network. *arXiv Preprint arXiv:2001.01100* (2020).

88   Regan, E. A. *et al.* Genetic epidemiology of COPD (COPDGene) study design. *COPD* **7**, 32-43 (2010).

89   Ashcroft, T., Simpson, J. M. & Timbrell, V. Simple method of estimating severity of pulmonary fibrosis on a numerical scale. *Journal of Clinical Pathology* **41**, 467-470 (1988).

90   Leek, J. T. *et al.* Tackling the widespread and critical impact of batch effects in high-throughput data. *Nature Reviews Genetics* **11**, 733-739 (2010).

91   Langer, S., Taubmann, O., Denzinger, F., Maier, A. & Mühlberg, A. DeepTechnome: Mitigating Unknown Bias in Deep Learning Based Assessment of CT Images. *arXiv Preprint arXiv:2205.13297* (2022).

92   Adams, J. E. Quantitative computed tomography. *European Journal of Radiology* **71**, 415-424 (2009).

93   Lee, D. C., Hoffmann, P. F., Kopperdahl, D. L. & Keaveny, T. M. Phantomless calibration of CT scans for measurement of BMD and bone strength—inter-operator reanalysis precision. *Bone* **103**, 325-333 (2017).

94   Traverso, A., Wee, L., Dekker, A. & Gillies, R. Repeatability and reproducibility of radiomic features: a systematic review. *International Journal of Radiation Oncology, Biology, Physics* **102**, 1143-1158 (2018).

95   Fortin, J.-P. *et al.* Removing inter-subject technical variability in magnetic resonance imaging studies. *NeuroImage* **132**, 198-212 (2016).

96   Mühlberg, A. *et al.* The Technome-A Predictive Internal Calibration Approach for Quantitative Imaging Biomarker Research. *Scientific Reports* **10,** 1103 (2020).

97   Muehlberg, A., Taubmann, O., Katzmann, A. & Suehling, M. Method for obtaining disease-related clinical information. *U.S. Patent Application No. 17/109,332*.

98   Sasaki, K. *et al.* The tumor burden score: a new "metro-ticket" prognostic tool for colorectal liver metastases based on tumor size and number of tumors. *Annals of Surgery* **267**, 132-141 (2018).

99   Chlebus, G. *et al.* Automatic liver tumor segmentation in CT with fully convolutional neural networks and object-based postprocessing. *Scientific Reports* **8**, 15497 (2018).

100   Bronstein, M. M., Bruna, J., LeCun, Y., Szlam, A. & Vandergheynst, P. Geometric deep learning: going beyond euclidean data. *IEEE Signal Processing Magazine* **34**, 18-42 (2017).

101   Weinstein, J. N. *et al.* The cancer genome atlas pan-cancer analysis project. *Nature Genetics* **45**, 1113-1120 (2013).

102   Katzmann, A. *et al.* Explaining clinical decision support systems in medical imaging using cycle-consistent activation maximization. *Neurocomputing* **458**, 141-156 (2021).

103   Goodfellow, I. *et al.* Generative adversarial networks. *Communications of the ACM* **63**, 139-144 (2020).

104   Bengio, Y., Louradour, J., Collobert, R. & Weston, J. Curriculum learning. in *2009 Proceedings of the 26th annual international Conference on Machine Learning,* 41-48.

105   Breiman, L. Random forests. *Machine Learning* **45**, 5-32 (2001).

106   Zhu, J.-Y., Park, T., Isola, P. & Efros, A. A. Unpaired image-to-image translation using cycle-consistent adversarial networks. in *Proceedings of the 2017 IEEE International Conference on Computer Vision.* 2223-2232 (IEEE).

107   Engstrom, C. M. *et al.* Segmentation of the quadratus lumborum muscle using statistical shape modeling. *Journal of Magnetic Resonance Imaging* **33**, 1422-1429 (2011).

108   Essafi, S. *et al.* Essafi, Salma, et al. Wavelet-driven knowledge-based MRI calf muscle segmentation. in *2009 IEEE International Symposium on Biomedical Imaging: From Nano to Macro.* 225-228 (IEEE).

109   Kamiya, N. *et al.* Automated segmentation of psoas major muscle in X-ray CT images by use of a shape model: preliminary study. *Radiological Physics and Technology* **5**, 5-14 (2012).

110   Ronneberger, O., Fischer P., & Brox T. "U-net: Convolutional networks for biomedical image segmentation." in *2015 Medical Image Computing and Computer-Assisted Intervention. Proceedings Part III 18*, 234-241 (Springer).

111   Mercuri, E. *et al.* Muscle MRI in inherited neuromuscular disorders: past, present, and future. *Journal of Magnetic Resonance Imaging: An Official Journal of the International Society for Magnetic Resonance in Medicine* **25**, 433-440 (2007).

112   Kang, Y., Engelke, K. & Kalender, W. A. Interactive 3D editing tools for image segmentation. *Medical Image Analysis* **8**, 35-46 (2004).

113   Mühlberg, A. *et al.* Three-dimensional distribution of muscle and adipose tissue of the thigh at CT: Association with acute hip fracture. *Radiology* **290**, 426-434 (2019).

114   Mühlberg, A. *et al.* SEMPAI: a Self-Enhancing Multi-Photon Artificial Intelligence for Prior-Informed Assessment of Muscle Function and Pathology. *Advanced Science* **10**, 2206319 (2023*).*

115   Tan, M. & Le, Q. EfficientNet: rethinking model scaling for convolutional neural networks. in *2019 International Conference on Machine Learning.* 6105-6114 (PMLR).

116   Deb, K., Pratap, A., Agarwal, S. & Meyarivan, T. A fast and elitist multiobjective genetic algorithm: NSGA-II. *IEEE Transactions on Evolutionary Computation* **6**, 182-197 (2002).

117 Parmentier, L., Nicol, O., Jourdan, L. & Kessaci, M.-E. TPOT-SH: A faster optimization algorithm to solve the AutoML problem on large datasets. in *2019 IEEE 31st International Conference on Tools with Artificial Intelligence.* 471-478 (IEEE).

118 Caruana, R. Multitask learning. *Machine Learning* **28**, 41-75 (1997).

119 Kendall, A., Gal, Y. & Cipolla, R. Multi-task learning using uncertainty to weigh losses for scene geometry and semantics. in *Proceedings of the 2018 IEEE Conference on Computer Vision and Pattern Recognition.* 7482-7491 (IEEE).

120 Lundberg, S. M. *et al.* From local explanations to global understanding with explainable AI for trees. *Nature Machine Intelligence* **2**, 56-67 (2020).

121 Lundberg, S. M. & Lee, S.-I. A unified approach to interpreting model predictions. *Advances in Neural Information Processing Systems* **30**, 4768-4777 (2017).

122 Lou, B. *et al.* An image-based deep learning framework for individualising radiotherapy dose: a retrospective analysis of outcome prediction. *The Lancet Digital Health* **1**, e136-e147 (2019)

123 Karniadakis, G. E. *et al.* Physics-informed machine learning. *Nature Reviews Physics* **3**, 422-440 (2021).

124 Rudin, C. Stop explaining black box machine learning models for high stakes decisions and use interpretable models instead. *Nature Machine Intelligence* **1**, 206-215 (2019).

125 Sutton, R. *The Bitter Lesson*, <https://www.cs.utexas.edu/~eunsol/courses/data/bitter_lesson.pdf> (2019).

126 Hosny, A., Aerts, H. J. & Mak, R. H. Handcrafted versus deep learning radiomics for prediction of cancer therapy response. *The Lancet Digital Health* **1**, e106-e107 (2019).

127 Jumper, J. *et al.* Highly accurate protein structure prediction with AlphaFold. *Nature* **596**, 583-589 (2021).

128 Zhang, B. H., Lemoine, B. & Mitchell, M. Mitigating unwanted biases with adversarial learning. in *Proceedings of the 2018 AAAI/ACM Conference on AI, Ethics, and Society.* 335-340.

129 Creswell, A. *et al.* Generative adversarial networks: an overview. *IEEE Signal Processing Magazine* **35**, 53-65 (2018).

130 Johansson, T. & Meier, P. A finite-element model for the mechanical analysis of skeletal muscles. *Journal of Theoretical Biology* **206**, 131-149 (2000).

131 Fuller, A., Fan, Z., Day, C. & Barlow, C. Digital twin: Enabling technologies, challenges and open research. *IEEE Access* **8**, 108952-108971 (2020).

132 Coorey, G. *et al.* The health digital twin to tackle cardiovascular disease—a review of an emerging interdisciplinary field. *NPJ Digital Medicine* **5**, 126 (2022).

5. Original Publications

Reproduced with permission from John Wiley and Sons, Springer Nature, the Public Library of Science (PLOS), and the Radiological Society of North America (RSNA).

Received: 23 September 2020 | Revised: 20 April 2021 | Accepted: 1 June 2021

DOI: 10.1002/mp.15049

RESEARCH ARTICLE

MEDICAL PHYSICS

Unraveling the interplay of image formation, data representation and learning in CT-based COPD phenotyping automation: The need for a meta-strategy

Alexander Mühlberg[1] | Rainer Kärgel[1] | Alexander Katzmann[1] | Felix Durlak[1] | Paul-Edouard Allard[2] | Jean-Baptiste Faivre[2] | Michael Sühling[1] | Martine Rémy-Jardin[2] | Oliver Taubmann[1]

[1]CT R&D Image Analytics, Siemens Healthineers, Forchheim, Germany

[2]Hospital Calmette, University Centre of Lille, Lille, France

Correspondence
Alexander Mühlberg, CT R&D Image Analytics, Siemens Healthineers, Forchheim, Germany.
Email: alexander-muehlberg@hotmail.com

Abstract

Purpose: In the literature on automated phenotyping of chronic obstructive pulmonary disease (COPD), there is a multitude of isolated classical machine learning and deep learning techniques, mostly investigating individual phenotypes, with small study cohorts and heterogeneous meta-parameters, e.g., different scan protocols or segmented regions. The objective is to compare the impact of different experimental setups, i.e., varying meta-parameters related to image formation and data representation, with the impact of the learning technique for subtyping automation for a variety of phenotypes. The identified associations of these parameters with automation performance and their interactions might be a first step towards a determination of optimal meta-parameters, i.e., a meta-strategy.

Methods: A clinical cohort of 981 patients (53.8 ± 15.1 years, 554 male) was examined. The inspiratory CT images were analyzed to automate the diagnosis of 13 COPD phenotypes given by two radiologists. A benchmark feature set that integrates many quantitative criteria was extracted from the lung and trained a variety of learning algorithms on the first 654 patients (two thirds) and the respective algorithm retrospectively assessed the remaining 327 patients (one third). The automation performance was evaluated by the area under the receiver operating characteristic curve (AUC). 1717 experiments were conducted with varying meta-parameters such as reconstruction kernel, segmented regions and input dimensionality, i.e., number of extracted features. The association of the meta-parameters with the automation performance was analyzed by multivariable general linear model decomposition of the automation performance in the contributions of meta-parameters and the learning technique.

Results: The automation performance varied strongly for varying meta-parameters. For emphysema-predominant phenotypes, an AUC of 93%–95% could be achieved for the best meta-configuration. The airways-predominant

Martine Rémy-Jardin and Oliver Taubmann share senior authorship (last author).

[Correction added on 29, Jul 2021 after initial online publication: the text 'Unraveling' has been added to the title.]

phenotypes led to a lower performance of 65%–85%, while smooth kernel configurations on average were unexpectedly superior to those with sharp kernels. The performance impact of meta-parameters, even that of often neglected ones like the missing-data imputation, was in general larger than that of the learning technique. Advanced learning techniques like 3D deep learning or automated machine learning yielded inferior automation performance for non-optimal meta-configurations in comparison to simple techniques with suitable meta-configurations. The best automation performance was achieved by a combination of modern learning techniques and a suitable meta-configuration.

Conclusions: Our results indicate that for COPD phenotype automation, study design parameters such as reconstruction kernel and the model input dimensionality should be adapted to the learning technique and may be more important than the technique itself. To achieve optimal automation and prediction results, the interaction between input those meta-parameters and the learning technique should be considered. This might be particularly relevant for the development of specific scan protocols for novel learning algorithms, and towards an understanding of good study design for automated phenotyping.

KEYWORDS
computed tomography, COPD, deep learning, image analysis, machine learning

1 | INTRODUCTION

Chronic obstructive pulmonary disease (COPD) is a leading cause of morbidity and mortality worldwide that induces an economic and social burden that is both substantial and increasing.[1] It represents an important public health challenge ranging from prevention, in the fields of smoking and environmental pollution, to optimized management of a multifaceted disease. Defined by an irreversible airflow obstruction assessed by spirometry (i.e., a fixed ratio of FEV1/FVC < 0.70, where FEV1 denotes the forced expiratory volume in one second and FVC the forced vital capacity), COPD is further characterized by the severity of airflow limitation with recommendations for an active search for common co-morbidities, such as cardiovascular disease, metabolic syndrome, osteoporosis, and lung cancer, that may interfere with patients' clinical presentation and adversely affect their outcome.[2]

Interpretation of chronic airflow limitation that characterizes COPD is not straightforward as it can be caused by a mixture of small-airway disease and parenchymal destruction, the relative contribution of which varies from person to person.[1] The identification of predominant structural changes is the foundation of computed tomography (CT)-based phenotyping of COPD, allowing identification of two main subgroups of diseases, i.e., emphysema-predominant (EPD) and airway-predominant diseases (APD).[3] Patients with EPD phenotype are associated with low body mass index (BMI),[4] severe dyspnea,[5] rapid decline in

FEV1,[6] and high respiratory-related mortality.[7] Patients with airway-predominant phenotype are associated with frequent chronic cough[8] and frequent exacerbations.[9] Patients with the mixed phenotype are associated with more severe dyspnea and more frequent hospitalizations.[10]

The automation of the COPD phenotype identification with CT is a long-standing research task. For the identification of emphysema and its subtypes, a variety of quantitative methods exist, starting with simple descriptions by density thresholds like the percentage of voxels below −950 HU (%LAA-950) as a surrogate for emphysema, continuing with methods using texture analysis.[3,11] For the APD phenotypes, parametric response mapping (PRM) was used to identify air trapping by registration of inspiratory and expiratory CT scans.[12] There are also approaches for bronchiolectasis[13] identification and automated tree-in-bud sign detection.[14] Regarding APD phenotypes, one should also mention machine learning (ML) methods to detect the phenotype of pulmonary micronodules, which represent early signs of smoking-related tissue alterations.[15]

The introduction of deep learning (DL), or more specifically convolutional neural networks, in the field of radiological image analysis[24] has led to development of newer methods for COPD phenotyping. For example, a 2D DL algorithm has been used for CT data to predict the COPD GOLD stage,[17] a score which is based on functional and clinical criteria.

In this context, a striking finding from the COPD-related literature is the lack of a holistic approach: Usually only

MEDICAL PHYSICS | 5181

TABLE 1 Exemplary selected studies for automated CT-based Chronic Obstructive Pulmonary Disease (COPD) phenotyping

CT Study	Cohort size	Phenotype(s)	Method	Features	Representation
Humphries[16]	9652	E (Fleischner)	2D DL	DL	Every 5 slices
Gonzalez[17]	7983	E (Fleischner)	2D DL	DL	4 slices
Monkam[15]	1010	Micronodules	2D DL	DL	Patches
Galban[12]	194	SAD	PRM	PRM	Whole lung
Bagci[14]	60	TIB	Manual tuning	Texture, shape	Whole lung
Park[18]	39	E (2 grades)	ML	Texture, shape	Preselected ROIs
Kim[19]	82	CLE, Bronchiol.	ML	Texture, shape	Preselected ROIs
Peng[20]	91	CLE (3 grades), PSE (2), PLE	2D DL	DL	Patches
Zulueta-Coarasa[21]	267	CLE, PSE, PLE	2D DL	DL	Patches
Hoff[22]	88	E, SAD, AT	PRM	PRM	Whole lung
Hosseini[23]	24	AT	~PRM	~PRM	Whole lung
Sorensen[11]	39	CLE, PLE, PSE	ML	Texture	Preselected ROIs

Abbreviations: aBWT, abnormal bronchial wall thickening; AT, air trapping.CLE, centrilobular emphysema; DL, deep learning; E, emphysema; ML, machine learning; PLE, panlobular emphysema; PRM, parameteric response mapping; PSE, paraseptal emphysema; ROI, region-of-interest; SAD, small airways disease; TIB, tree-in-bud.

individual aspects, phenotypes or methods, are examined. As can be seen from Table 1, proprietary methods were developed to automate the detection of one or few phenotypes, often for small study cohorts, which may lead to results that possibly do not generalize across biological variability. Another factor is that scan protocols differ between publications. Also, the data representation fed into machine learning (ML) or DL often appears to be somewhat arbitrary, i.e., which segmentations, slices or patches are selected varies between studies. The selected learning technique is usually highlighted as the most important factor. Those decisions regarding the experimental setup and used learning techniques do not always appear systematic rendering the results difficult to compare. These problems are likewise known for computational COPD phenotyping, where "[...] studies used different methods and variables, making it challenging to synthesize and interpret this literature".[25]

We therefore employ a variety of learning techniques and vary, e.g., the segmented regions used as input and the reconstruction kernel, for a large number of phenotypes and a large COPD study cohort. This holistic approach should allow us to determine associations of the meta-parameters, i.e., experimental setups as described above, with the automation performance and subsequently identify a meta-strategy, i.e., how to determine optimal meta-parameters.

Our main objective is to establish the first step towards a meta-strategy based on a large clinical cohort and an exhaustive phenotyping, to understand the influence of meta-parameters related to image formation, data representation, and learning on the automation/prediction performance and to identify an optimal configuration of the meta-parameters. Our second objective is to provide automation results for a larger cohort

given an optimized meta-configuration using only a simple unified set of features for a variety of phenotypes and compare them to results with current advanced approaches found in the literature that were mostly engineered for single phenotypes.

2 | MATERIALS AND METHODS

The study protocol was approved by our institutional Ethics Committee with waiver of informed consent as CT investigations were part of routine practice. Their retrospective reading did not require patient informed consent.

2.1 | Study population

A total of 981 patients were collected at Hospital Calmette, University Centre of Lille, France, in the period from October 2016 to October 2017 with either of the criteria (a) normal chest CT examination; (b) presence of CT features of COPD of variable severity; in both categories, the presence of minor additional changes (i.e., few indeterminate lung micronodules, intrapulmonary lymph node(s) and/or a few areas of linear atelectasis) was considered as acceptable. Characteristics of the final study cohort are shown in Table 2.

2.2 | Image annotation and inter-reader variance

All examinations had been obtained on a 3rd-generation dual-source CT system (SOMATOM Force;

TABLE 2 Patient characteristics and scan and reconstruction parameters of the analyzed study cohort

Variable	Value
Patients	981
Pack years	20.4 ± 24
Age	53.8 ± 15.1
Male sex	554
Nonsmoker	292
Kernel	"smooth": Bl57; "sharp": Br31
Slice thickness/distance	1 mm/1 mm

TABLE 3 11 Chronic Obstructive Pulmonary Disease subtypes and their assignment to emphysema—or airways—predominant COPD

COPD Type	Labels
Emphysema-predominant (EPD)	**Emphysema (E)**
	Centrilobular Emphysema (CLE)
	Panlobular Emphysema (PLE)
	Paraseptal Emphysema (PSE)
	Bullae
Airways-predominant (APD)	**Small-airways disease (SAD)**
	Abnormal bronchial thickening (aBWT)
	Air Trapping (AT)
	Tree-in-Bud (TIB)
	Bronchiolectasis
	Micronodules

Note: EPD and APD are also analyzed as phenotypes in a broader definition, resulting in 13 analyzed phenotypes

Siemens Healthineers, Forchheim, Germany) with single-energy, high-pitch scanning mode, after deep inspiration; the scanning parameters ensured a radiation dose adapted to the patient's body morphotype with a kilovoltage ranging between 80 and 120 kVp and adapted mAs (dose length product ranging between 80 and 250 mGy·cm). The data sets generated lung and mediastinal images, reconstructed with sharp high-spatial (Bl57) and smooth (Br36) kernels, respectively.

Expiratory acquisitions were not systematically obtained. The list of studied phenotypes is given in Table 3. All selected examinations were labeled after a consensus reading between two radiologists, the senior radiologist in charge of case selection (MRJ) and a junior radiologist (PEA) with more than 4 years of experience in CT, which established the ground truth labels. On each examination, they rated the presence of (a) abnormal bronchial wall thickening of central airways; (b) CT indications of small-airway disease (SAD), including abnormal thickening of

bronchiolar walls (aBWT)—either of normal diameter or dilated—bronchiolectasis, tree-in-bud pattern (TIB), and/or ill-defined micronodules; air trapping (AT) could only be investigated on examinations that had included expiratory acquisitions; emphysematous subtypes centrilobular (CLE), paraseptal (PSE), and panlobular (PLE), based on the Fleischner Society recommendations.[3]

2.3 | Segmentations and feature extraction

An automated segmentation algorithm[26] was applied on the images to separate the lung into two left (upper, lower) and three right (upper, middle, lower) lobes. For an automated segmentation of the inner and outer airway walls to extract related measurements, the method described by Odry et al. was used.[27] Lung vessels were automatically segmented by the algorithm of Kaftan et al.[28] A variety of non-lung regions-of-interest (NL-ROIs) in the heart, aorta, adipose tissue, liver, spleen, air, trachea were determined by the automated segmentation algorithm by Seifert et al.[29]

Quantitative imaging features were engineered for COPD phenotyping and are used as a benchmark to assess the impact of meta-parameters. Those features can be classified in 4 groups, (i) density-related measurements in the parenchyma such as threshold-based %LAA-950 and %LAA-910[30] that measure the volume percentage of lung voxels below −950 and −910 HU, and texture features to quantify the extent of different types of emphysema, (ii) airways features such as wall thickness, (iii) lung vessel features and (iv) features intended to detect ill-defined micronodules. The latter abnormalities were considered owing to the relationships between tobacco-related inflammatory changes at the level of distal airways and development of emphysema.[31] Descriptions, references and the rationale of our feature library can be found in the Section S1. For the NL-ROIs, the PyRadiomics[32] features were calculated.

2.4 | COPD meta-parameter analytics

Our approach compared to the solutions in the literature of Table 1 is shown in Table 4. A flowchart describing the general evaluation setup is found in Figure 1.

To determine the automation performance, we employed a temporal split, i.e., two thirds of the patients with earlier scan dates were used for training and the other third was used for evaluating the automation method (Figure 1). The automation performance for each phenotype is given by the area under the receiver operating characteristic curve (AUC) and analyzed

TABLE 4 Our COPD phenotyping approach, which investigates the impact of different learning techniques and experimental setups for a variety of phenotypes

Cohort size	Phenotypes	Methods	Features	Representation
981 (654 / 327)	EPD (208 / 103) E (262 / 114) CLE (245 / 100) PLE (34 / 34) PSE (182 / 80) Bullae (70 / 20) APD (229 / 116) SAD (177 / 93) aBWT (307 / 153) AT (101 / 43) TIB (22 / 13) Bronch. (18 / 15) Micron. (97 / 52)	Statistical Pipeline/ML: XGB, RF, /AutoML: TPOT 3D DL: VoxResNet (E only)	Density, Texture, Airways, Vessels DL	Whole Lung, Lung Lobes, airways, vessels, NL-ROIs

Note: In parentheses, the number of cases labeled positive for each phenotype in our (training/test) data sets is stated.

(Auto)ML, (automated) machine learning; aBWT, abnormal bronchial wall thickening; APD, airways-predominant; AT, air trapping; bronch., bronchiolitis; CLE, centrilobular emphysema; DL, deep learning; E, emphysema; EPD, emphysema-predominant; micron., micronodules.NL-ROIs, nonlung regions-of-interest; PLE, panlobular emphysema; PSE, paraseptal emphysema; SAD, small airways disease; TIB, tree-in-bud.

FIGURE 1 Evaluation setup including data collection, pre-training of the three-dimensional deep learning algorithm, temporal data split and application for the phenotype detection [Color figure can be viewed at wileyonlinelibrary.com]

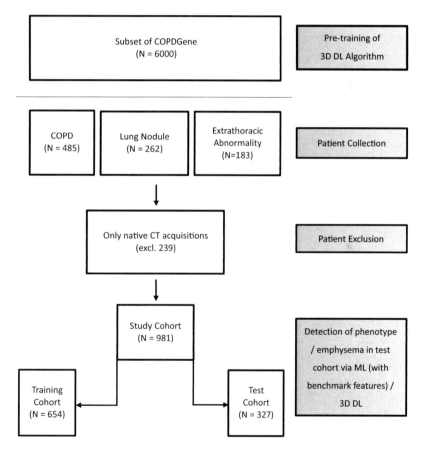

for different experimental setups. The following meta-parameters are varied:

(i) The learning technique. For ML techniques we use extreme gradient boosting (XGB),[33] random forests[34] without and with hyperparameter optimization (RF and RF_opt), and an automated machine learning (AutoML) algorithm, namely the tree-based TPOT-SH algorithm.[35,36] These methods were selected since they represent the state of the art in conventional machine learning methods. Except for RF (which is complemented by RF_opt), all methods employ some form of hyperparameter optimization, which we consider particularly important to ensure that the reliability of our findings is not impeded by arbitrary, suboptimal choices for model-specific settings. Additionally, the automation performance of a statistical approach[37] is analyzed. Our statistical pipeline selects significant features (T-test or Wilcoxon rank-sum test; $p < 0.05$) after Benjamini-Hochberg multiple testing correction with subsequent mRMR[38] feature selection to pre-select up to 10 relevant and non-redundant features and an Akaike Information Criterion best subset selection, all performed on the training data set. The motivation behind feature selection is to ensure that models are not "enriched" with too many irrelevant or uninformative input features, which may deteriorate their ability to learn. The identified subset of features is then used for a simple multiple logistic regression for the training data and the model is used to predict the presence of the phenotype for the test data. In addition to these feature-based approaches, a specifically developed 3D DL algorithm[39] is used to automate the detection of emphysema based on the segmentation of the entire lung (more information can be found in the Section S2). This 3D DL algorithm was pretrained on COPDGene[30] where it achieved an AUC of 81% for COPD detection based on GOLD stage criteria, comparable to the state-of-the-art 2D DL method of Gonzalez et al. that achieved 86%.[17]

(ii) The choice of segmented regions and/or used features used as input for learning is always a tradeoff between focus and context (Figure 2), where a larger input dimensionality may result in overfitting while focusing on a smaller region or fewer features may miss important information. We therefore compare whether it is advantageous to calculate the features for each lobe of the lungs separately and concatenate them afterwards, or to calculate them for the whole lung region instead. We also analyze the impact of incorporating additional features from NL-ROIs. These features may

be helpful for two reasons, either by providing relevant biological information or, indirectly, by internally calibrating the measurements within the lung by features in NL-ROIs.[40,41] For these meta-configurations the mRMR algorithm is always used to slightly reduce the impact of overfitting.

(iii) It is assumed that APD phenotypes should be analyzed with a sharp reconstruction kernel to prevent overestimation of airway wall sizes,[42] while for EPD a smooth kernel is to be preferred as it leads to higher correlation of some established emphysema quantifications with histological COPD measurement.[43] We analyze whether this also applies to learning-based phenotyping automation by varying the selected reconstruction kernel.

(iv–vi) Finally, three often neglected meta-parameters are analyzed. First, we investigate the benefit of incorporating standard clinical (age, sex) and scan protocol parameters (kVp, protocol identifier: mixed image/single energy) in the training process. Clinical parameters are often correlated with the investigated diagnoses and may thus yield complementary information, and providing scan protocol parameters could allow models to more readily adapt to different image characteristics. Second, how to deal with missing feature values, as not all airway features may be computable in each patient. This is due to the fact that they are calculated for each automatically detected airway segment and subsequently averaged over the generations, i.e., over all segments with the same depth in the bronchial tree, whereby depth refers to the number of bifurcations on the direct path from the trachea to a segment. Depending on the image quality, fine vessels of higher generations may not be

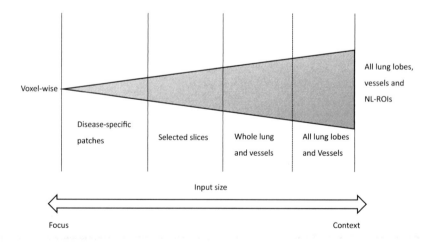

FIGURE 2 Focus vs. context-tradeoff for artificial intelligence research in Chronic Obstructive Pulmonary Disease phenotyping. The input size and input dimensionality increases from left to right, indicated by solid lines. The dashed line separates the representations that share the same input size but have different input dimensionality, as more features can trivially be extracted, when all lung lobes are analyzed individually. While a more context-preserving input may confuse an AI system with unnecessary information and lead to overfitting, a more focused input is limited as it may miss importance contextual information or in general information which could be relevant to the learning process [Color figure can be viewed at wileyonlinelibrary.com]

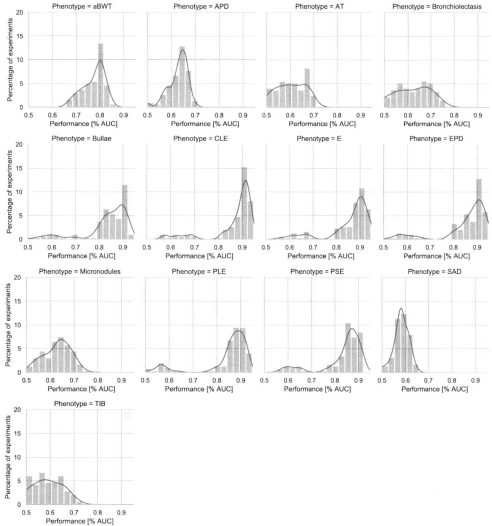

FIGURE 3 Variation of automation performance [% AUC] of different phenotypes due to variation of all meta-parameters. y-axis: percentage of experiments on all experiments, x-axis: automation performance distribution. Barplots are shown with kernel density estimate and normalized to sum up to 100% [Color figure can be viewed at wileyonlinelibrary.com]

detected by the method. In that case, no features can be computed for these generations. We therefore test two different imputation thresholds: If more than 15% or 50% of the feature values are missing, the feature is discarded, otherwise a mean imputation is applied. Third, the benefit of the mRMR feature selection as a dimensionality reduction of the ML training upstream of the ML training is evaluated.

The association of meta-parameters with the automation performance is then analyzed by a multivariable general linear model. The automation performance P for a phenotype is decomposed into the base performance, or intercept, p_0 and the influence of all meta-parameters,

$$P_i = p_0 + \sum_{j \in M} p_j C_{ij} + \varepsilon_i,$$

with experiment i of the set of experiments, meta-parameter j of the set of meta-parameters M and configuration $C_{ij} \in 0, 1$, where $C_{ij} = 1$ if the respective meta-parameter configuration j is set for patient i, and the error ε_i. The model is termed "decomposition model". The fitted coefficients $\hat{p}_j$ of this model, which represent the estimated performance gain or loss for a specific configuration, e.g., the performance gain or loss when the kernel is changed from smooth to sharp, are fed into a clustermap with a dendrogram to identify phenotypes that prefer similar meta-configurations. For each phenotype, the base performance $\hat{p}_0$ and the highest achievable automation performance with an optimal meta-configuration are shown. This base performance is the model-estimated performance of the configuration with all meta-parameters $C_{ij} = 0$, which corresponds to

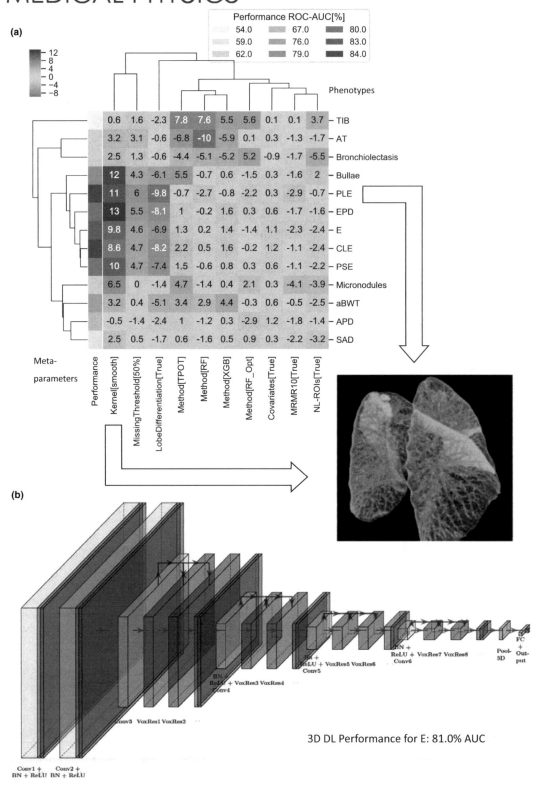

(a)

(b)

3D DL Performance for E: 81.0% AUC

■ Convolutional layer ■ Batch normalization ■ Activation layer ■ VoxRes module

FIGURE 4 (a) Decomposition model for the association of automation performance variation P with meta-parameter variation. The color-coded (blue to red) values indicate the performance gain or loss $\hat{p}_j$ for the meta-parameter j when the corresponding meta-configuration, e.g., smooth kernel, was set. Green indicates the intercept, i.e., the base performance $\hat{p}_0$ to which the loss or gain is added to obtain the performance estimate. For example, the choice of a smooth kernel had the greatest positive impact on automation performance and the influence of meta-configuration was highest for PLE, where the difference between a good and a bad meta-configuration adds up to an automation performance difference of ca. 35% AUC (indicated by arrows). Note: for the meta-parameter "Method", the statistical pipeline is the base configuration and therefore omitted. (b) The architecture the 3D DL algorithm applied on the whole lung is shown, which resulted in an automation performance for emphysema of 81.0% [Color figure can be viewed at wileyonlinelibrary.com]

TABLE 5 Achievable automation performance for a variety of phenotypes assessed with the optimal meta-configuration. The cutoff for dichotomisation is determined by Youden's index

Phenotype	AUC	CI lower	CI upper	Sens	Spec	Acc	F1	Inform	Mark	Matthews
EPD	0.95	0.92	0.97	0.86	0.90	0.89	0.82	0.76	0.73	0.74
E	0.95	0.93	0.97	0.89	0.89	0.89	0.85	0.78	0.75	0.76
CLE	0.95	0.92	0.97	0.91	0.85	0.87	0.80	0.76	0.68	0.72
PLE	0.94	0.89	0.97	0.88	0.87	0.87	0.58	0.75	0.42	0.56
PSE	0.93	0.90	0.95	0.89	0.85	0.86	0.76	0.74	0.62	0.68
Bullae	0.93	0.89	0.97	0.95	0.82	0.83	0.40	0.77	0.26	0.44
APD	0.71	0.64	0.76	0.71	0.63	0.66	0.59	0.34	0.31	0.32
SAD	0.65	0.58	0.71	0.63	0.65	0.64	0.49	0.27	0.23	0.25
AT	0.70	0.63	0.78	0.88	0.46	0.52	0.33	0.34	0.17	0.24
Bronchiol.	0.75	0.57	0.89	0.65	0.82	0.81	0.23	0.47	0.12	0.24
TIB	0.73	0.59	0.85	0.69	0.74	0.74	0.17	0.43	0.08	0.18
aBWT	0.85	0.81	0.89	0.83	0.73	0.78	0.78	0.57	0.57	0.57
Micronod.	0.75	0.68	0.81	0.78	0.66	0.68	0.43	0.44	0.24	0.32

CI lower, upper = lower or upper confidence interval, estimated by means of bootstrapping.[48] Sens = Sensitivity, Spec = Specifity, Acc = Accuracy, Inform = Informedness, Mark = Markedness, Matthews = Matthews Correlation Coefficient.

TABLE 6 Optimal meta-configuration for each phenotype. Meta-configuration using NL-ROIs were never optimal and this column is therefore omitted

Phenotype	Covariates	Kernel	Lobe Differentiation	mRMR	Method	ImputationThreshold
EPD	False	smooth	True	False	XGB	15.00
E	True	smooth	True	False	XGB	50.00
CLE	True	sharp	False	False	XGB	15.00
PLE	True	smooth	False	False	TPOT	50.0
PSE	True	smooth	True	False	TPOT	50.0
Bullae	True	smooth	True	False	TPOT	50.0
SAD	False	soft	False	True	RF_opt	15.00
APD	False	soft	False	False	XGB	50.00
AT	True	sharp	True	False	RF_opt	50.00
Bronchiolect.	False	sharp	False	False	RF_opt	50.00
TIB	False	smooth	True	False	RF	15.00
aBWT	False	smooth	True	False	XGB	50.00
Micronodules	False	smooth	False	False	RF_opt	50.00

a sharp reconstruction kernel, the 15% imputation threshold, the statistical prediction pipeline and the exclusion of the NL-ROIs, lobe differentiation, covariates and mRMR feature selection. For the best configuration, further performance metrics are calculated to enable a comparison with results in the literature.

3 | RESULTS

The variation of automation performance induced by meta-parameter variation is shown in Figure 3, the decomposition model in Figure 4a. The best performance for each phenotype can be found in Table 5, the corresponding meta-configuration in Table 6.

The performance for EPD phenotypes is generally higher than for APD phenotypes. For EPD and its subtypes, the best automation performance is very high with AUCs between 93 and 95%. The decomposition model shows that the influence of the kernel choice is dominant in this group with a gain of 8.6 to 13% AUC when a smooth kernel is selected. Another important factor is to set a rather tolerant imputation threshold, which discards features only if 50% of their values are missing, with an improvement of 4.3%–6.0% (Figure 4a), which indicates that airways features might also be important for EPD subtype automation (cf. 2.D. (v)). Dividing the lung into lobes and calculating the features for all lobes separately reduces the automation performance in the range of −6.1% to −9.8%. In the

decomposition model, ML was in general not superior to our statistical pipeline for EPD phenotype automation averaged over all meta-configurations. However, the best meta-configuration was always found with (Auto)ML (Table 6) and employed either XGB with hyperparameter optimization or TPOT. The 3D DL model trained on the whole lung yields only a moderate performance of 81% AUC for the automated detection of emphysema (Figure 4b). This is in contrast to 95% AUC achieved with conventional ML with a suitable meta-configuration. Additionally, the incorporation of NL-ROIs led to deteriorated performance, with exception of the EPD subtype Bullae and the APD subtype TIB. As can be seen in Figure 5, the performance distributions for (Auto)ML with hyperparameter optimization and the

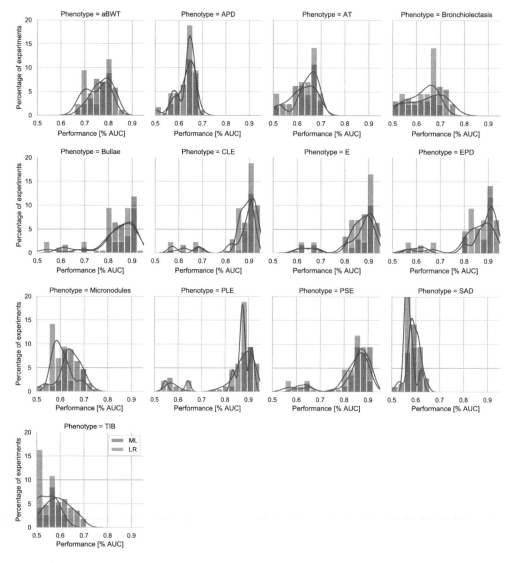

FIGURE 5 Automation performance [% AUC] with a statistical pipeline and logistic regression (LR, blue) and (Auto)ML with hyperparameter optimization (ML, red) for the 13 phenotypes. y-axis: percentage of experiments on all experiments, x-axis: automation performance [Color figure can be viewed at wileyonlinelibrary.com]

statistical pipeline are in general not significantly different (t test or Wilcoxon rank-sum test; $p > 0.05$), except for TIB and micronodules ($p < 0.01$). The dendrogram in Figure 4a shows that EPD phenotypes prefer similar meta-configurations while the meta-strategy for APD phenotypes is not that obvious.

For the APD subtypes, aBWT in particular can be automated with an AUC of 85% for an optimal meta-configuration. On the other hand, the automatic detection of SAD only achieved a performance of 65%. The remaining APD phenotypes achieve performances of 70%–75% AUC. Interestingly, also APD phenotypes were better automated with a smooth kernel, since only one of the seven APD phenotypes showed a higher performance for a sharp kernel when looking at the decomposition model, two when looking at the highest achievable automation performance. The phenotypes TIB, AT and bronchiolectasis benefit most from the selection of an appropriate learning technique, but its impact was never larger than the combined impact of all other meta-parameters. For APD phenotypes, the best performances are achieved by ML with hyperparameter optimization in six of seven cases, although the decomposition model does not confirm this trend.

4 | DISCUSSION

For this study, we have conducted 1717 experiments to understand the association of meta-parameters with automation performance and to identify optimal meta-configurations for COPD phenotyping. Our results raise questions as to whether the choice and fine-tuning of the learning technique is really appropriate without also optimizing meta-parameters such as reconstruction kernel or segmented regions, which often appear to be chosen arbitrarily.

If we compare our results with the most modern methods, it is noticeable that we were able to achieve comparatively good results when a suitable meta-configuration is selected, even with rather simple learning techniques and a uniform set of features for all phenotypes. Sörensen et al.[11] achieved an accuracy of 95% with specifically designed texture features for 39 patients, local binary patterns, and patches. Peng et al.[20] achieved 93% accuracy for PLE, PSE and CLE with a DL-based approach and image patches explicitly refined for this issue in a cohort of 91 patients. We achieve similar AUCs of 92%–95% and accuracies of 86%–89%—partly even with a simple logistic regression—in a cohort of 981 patients.

In our study, the impact of the reconstruction kernel on automation performance was larger than the learning technique. It could also be observed that although the sharp kernel may be marginally better for the automatic detection of APD, the smooth kernel is the better configuration for the automation of phenotypes

belonging to this class—which is contrary to the recommended setting for quantification of APD phenotypes.[42,43] Apparently, there might be other optimal meta-configurations for learning tasks than for quantification. Similar considerations may also be of interest in light of recent work that aims to increase the resolution of or reduce noise in CT images using learning-based methods,[44-46] where the benefit of such enhancements may be highly task-dependent as well.

For AT, the best result we are aware of is the work of Maansor et al.[47] for 51 radiographs, where DL achieved an F1-score of 77%. Our results on the whole lung yield only 33% F1-score, which is especially interesting since Maansor et al. work with lower input dimensionality. This deterioration of performance with a larger input dimensionality is also reflected in our 3D DL approach for the whole lung, which yields inferior performance than the feature-based, i.e., lower dimensionality, ML variants. Apparently, the input dimensionality is too large for the amount of data available, although our 3D DL network was pre-trained on 6000 patients of the COPDGene cohort and achieved results comparable to the state-of-the-art solution on that dataset. Also, the extraction of features from different lung lobes and the NL-ROIs did often not yield an improvement of the automation performance, but rather a deterioration. Even an exhaustive AutoML optimization did not yield good results for large input dimensionality, although the technique integrates a variety of state-of-the-art feature selection techniques. In the trade-off between focus and context it seems important to reduce the input dimensionality beforehand. In recent research by Humphries et al.[16] on the DL-based automation of GOLD stage scoring for emphysema, a "moderate" performance was achieved ("Weighted K Statistic comparing visual and deep learning scores was moderate"), while showing that automation facilitates correlation with outcome parameters like mortality. In the older study of Gonzalez et al.,[17] the 2D DL algorithm yielded a concordance index of 0.86 for COPD prediction based on the GOLD stage with only 4 selected slices over the whole lung. Based on our analyses, this could possibly be due to the choice of a more suitable input dimensionality in the approach of Gonzalez et al., that selected 1 axial, 2 sagittal and 1 coronal slice, instead of 25 axial slices over the whole lung, although the DL network capacity was lower. The optimized input selection, e.g., a different slice sampling strategy, could possibly have a similar effect as a fine-tuning of the architecture. Monkam et al. offered similar considerations when they analyzed the influence of patch sizes on the automation performance[15] for micronodules.

Even often neglected factors like the employed data imputation strategy may have a larger influence on the automation performance than the learning technique. Often, meta-configurations are only given in small print, if at all. However, the interaction of meta-parameters and the learning technique should be analyzed systematically. It appears that reconstruction parameters, input representation and dimensionality, i.e., the meta-strategy, should in general be optimized simultaneously with the learning technique as isolated optimization of one of these factors might lead to non-optimal solutions.

Our study has a number of limitations. First of all, we had to decide on a uniform exemplary set of benchmark features for the COPD phenotypes. The implementation of all features found in the literature would hardly have been possible also due to partially missing descriptions of these features. Second, we have not extended our analyses for APD to registered expiratory and inspiratory scans and methods like PRM.[12] The general automation performance for APD subtypes and especially SAD would probably have been higher with additional expiratory scans. However, expiratory scans were not systematically available in this clinical routine dataset and we assume that the importance of the meta-parameters would have been similar relative to the learning method. Third, it would also have been beneficial to vary the data input fed into DL, i.e., the data representation or slice sampling, as well as train DL models for all of the given tasks. The time and resources required to perform a comparable wealth of experiments with DL models proved prohibitive and, given the reliance of convolutional architectures on input images instead of preextracted features, would also require substantially different types of experiments. We consider this to be beyond the scope of this study and plan to address this topic in future work. Nevertheless, we believe the demonstrated comparison to a state-of-the-art DL model herein serves as an important point of reference and provides some intuition that such models do not effortlessly achieve superior and more robust performance that alleviates the concern to choose the best-suited image formation and data representation.

5 | CONCLUSIONS

The learning technique does not always have to be the decisive factor for good predictive/automation performance. In COPD phenotyping automation, and in particular when using complex learning techniques, it may be more important to identify suitable meta-configurations and especially adapt the input dimensionality, e.g., segmentation, slice or feature set, and data representation, e.g., kernel, to the technique used. The choice of these meta-configurations must also be optimized differently for learning than for radiological reading or simple quantifications. We therefore recommend large-scale studies, which acquire knowledge about suitable meta-configurations together with the learning techniques used, in order to systematically identify a meta-strategy and subsequently optimize scan protocols, image processing and learning technique simultaneously.

CONFLICTS OF INTEREST

A.M., R.K., A.K., F.D., M.S. and O.T. are employees of Siemens Healthineers. Academic radiologists had full control over the data and no product or commercial software was tested in the current study.

DATA AVAILABILITY STATEMENT

The data that support the findings of this study are available from the corresponding author upon reasonable request.

REFERENCES

1. Singh D, Agusti A, Anzueto A, et al. Global strategy for the diagnosis, management, and prevention of chronic obstructive lung disease: the GOLD science committee report 2019. *Eur Respir J.* 2019;53(5):1900164. https://doi.org/10.1183/13993003.00164-2019
2. Mannino DM, Thorn D, Swensen A, Holguin F. Prevalence and outcomes of diabetes, hypertension and cardiovascular disease in COPD. *Eur Respir J.* 2008;32(4):962–969. https://doi.org/10.1183/09031936.00012408.
3. Lynch DA, Austin JHM, Hogg JC, et al. CT-definable subtypes of chronic obstructive pulmonary disease: a statement of the Fleischner Society. *Radiology.* 2015;277(1):192-205.
4. Ogawa E, Nakano Y, Ohara T, et al. Body mass index in male patients with COPD: correlation with low attenuation areas on CT. *Thorax.* 2009;64(1):20-25.
5. Boschetto P, Quintavalle S, Zeni E, et al. Association between markers of emphysema and more severe chronic obstructive pulmonary disease. *Thorax.* 2006;61(12):1037-1042.
6. Vestbo J, Edwards LD, Scanlon PD, et al. Changes in forced expiratory volume in 1 second over time in COPD. *N Engl J Med.* 2011;365(13):1184-1192.
7. Haruna A, Muro S, Nakano Y, et al. CT scan findings of emphysema predict mortality in COPD. *Chest.* 2010;138(3):635-640.
8. Grydeland TB, Dirksen A, Coxson HO, et al. Quantitative computed tomography measures of emphysema and airway wall thickness are related to respiratory symptoms. *Am J Respir Crit Care Med.* 2010;181(4):353-359.
9. Han MK, Kazerooni EA, Lynch DA, et al. Chronic obstructive pulmonary disease exacerbations in the COPDGene study: associated radiologic phenotypes. *Radiology.* 2011;261(1):274-282.
10. Van Tho N, Ogawa E, Trang LTH, et al. A mixed phenotype of airway wall thickening and emphysema is associated with dyspnea and hospitalization for chronic obstructive pulmonary disease. *Ann Am Thorac Soc.* 2015;12(7):988-996.
11. Sorensen L, Shaker SB, De Bruijne M. Quantitative analysis of pulmonary emphysema using local binary patterns. *IEEE Trans Med Imaging.* 2010;29(2):559-569.
12. Galbán CJ, Han MK, Boes JL, et al. Computed tomography–based biomarker provides unique signature for diagnosis of COPD phenotypes and disease progression. *Nat Med.* 2012;18(11):1711.
13. Prasad M, Sowmya A, Wilson P. Automatic detection of bronchial dilatation in HRCT lung images. *J Digit Imaging.* 2008;21(Suppl 1):S148-S163. https://doi.org/10.1007/s10278-008-9113-4.
14. Bagci U, Yao J, Wu A, et al. Automatic detection and quantification of tree-in-bud (TIB) opacities from CT scans. *IEEE Trans Biomed Eng.* 2012;59(6):1620-1632.

15. Monkam P, Qi S, Xu M, Han F, Zhao X, Qian W. CNN models discriminating between pulmonary micro-nodules and non-nodules from CT images. *Biomed Eng Online.* 2018;17(1):96.

16. Humphries SM, Notary AM, Centeno JP, et al. Deep learning enables automatic classification of emphysema pattern at CT. *Radiology.* 2020;294(2).434-444. https.//doi.org/10.1148/radiol.2019191022.

17. González G, Ash SY, Vegas-Sánchez-Ferrero G, et al. Disease staging and prognosis in smokers using deep learning in chest computed tomography. *Am J Respir Crit Care Med.* 2018;197(2):193-203.

18. Park YS, Seo JB, Kim N, et al. Texture-based quantification of pulmonary emphysema on high-resolution computed tomography: comparison with density-based quantification and correlation with pulmonary function test. *Invest Radiol.* 2008;43(6):395-402. https://doi.org/10.1097/RLI.0b013e31816901c7.

19. Kim N, Seo JB, Lee Y, Lee JG, Kim SS, Kang S-H. Development of an automatic classification system for differentiation of obstructive lung disease using HRCT. *J Digit Imaging.* 2009;22(2):136-148.

20. Peng L, Lin L, Hu H, et al. Classification of pulmonary emphysema in ct images based on multi-scale deep convolutional neural networks. IEEE. 2018;3119-3123.

21. Zulueta-Coarasa T, Kurugol S, Ross JC, Washko GG, Estépar RSJ. Emphysema classification based on embedded probabilistic PCA. IEEE. 2013;3969-3972.

22. Hoff BA, Pompe E, Galbán S, et al. CT-based local distribution metric improves characterization of COPD. *Sci Rep.* 2017;7(1):2999. https://doi.org/10.1038/s41598-017-02871-1

23. Hosseini MP, Soltanian-Zadeh H, Akhlaghpoor S, Behrad A. A new scheme for evaluation of air-trapping in CT images. IEEE. 2010;1-4.

24. Sahiner B, Pezeshk A, Hadjiiski LM, et al. Deep learning in medical imaging and radiation therapy. *Med Phys.* 2019;46(1):e1-e36.

25. Castaldi PJ, Boueiz A, Yun J, et al. Machine learning characterization of COPD Subtypes: insights from the COPDGene study. *Chest.* 2019;157:114-157.

26. Odry B, Steininger P, Zhang L, et al. Pulmonary lobe segmentation using the thin plate spline (TPS) with the help of the fissure localization areas. International Society for Optics and Photonics. 2013:86690X.

27. Odry BL, Kiraly AP, Novak CL, Naidich DP, Lerallut J-F. Automated airway evaluation system for multi-slice computed tomography using airway lumen diameter, airway wall thickness and broncho-arterial ratio. International Society for Optics and Photonics. 2006:61430Q.

28. Kaftan JN, Kiraly AP, Bakai A, Das M, Novak CL, Aach T. Fuzzy pulmonary vessel segmentation in contrast enhanced CT data. International Society for Optics and Photonics; 2008:69141Q.

29. Seifert S, Barbu A, Zhou SK, et al. Hierarchical parsing and semantic navigation of full body CT data. International Society for Optics and Photonics; 2009:725902.

30. Regan EA, Hokanson JE, Murphy JR, et al. Genetic epidemiology of COPD (COPDGene) study design. *COPD.* 2010;7(1):32-43. https://doi.org/10.3109/15412550903499522.

31. Remy-Jardin M, Edme J-L, Boulenguez C, Remy J, Mastora I, Sobaszek A. Longitudinal follow-up study of smoker's lung with thin-section CT in correlation with pulmonary function tests. *Radiology.* 2002;222(1):261-270.

32. van Griethuysen JJM, Fedorov A, Parmar C, et al. Computational radiomics system to decode the radiographic phenotype. *Can Res.* 2017;77(21):e104-e107.

33. Chen T, Guestrin C. Xgboost: a scalable tree boosting system; 2016:785-794.

34. Breiman L. Random forests. *Mach Learn.* 2001;45(1):5-32.

35. Olson RS, Moore JH. TPOT. A tree-based pipeline optimization tool for automating machine learning. Automated Machine Learning. Springer. 2019;151-160.

36. Parmentier L, Nicol O, Jourdan L, Kessaci M-E. TPOT-SH: a Faster Optimization Algorithm to Solve the AutoML Problem on Large Datasets. IEEE. 2019;471-478.

37. Mühlberg A, Museyko O, Bousson V, Pottecher P, Laredo J-D, Engelke K. Three-dimensional distribution of muscle and adipose tissue of the thigh at CT: Association with acute hip fracture. *Radiology.* 2019;290(2):426-434.

38. Ding C, Peng H. Minimum redundancy feature selection from microarray gene expression data. *J Bioinform Comput Biol.* 2005;3(02):185-205.

39. Ahmed J, Vesal S, Durlak F, et al. COPD Classification in CT Images Using a 3D Convolutional Neural Network. 2020. *arXiv preprint arXiv:200101100.*

40. Mühlberg A, Katzmann A, Heinemann V, et al. The technome-a predictive internal calibration approach for quantitative imaging biomarker research. *Sci Rep.* 2020;10:1-5.

41. Leek JT, Scharpf RB, Bravo HC, et al. Tackling the widespread and critical impact of batch effects in high-throughput data. *Nat Rev Genet.* 2010;11(10):733-739.

42. Rodriguez A, Ranallo F, Judy P, Gierada D, Fain S. CT reconstruction techniques for improved accuracy of lung CT airway measurement. *Med Phys.* 2014;41(11):111911.

43. Gierada DS, Bierhals AJ, Choong CK, et al. Effects of CT section thickness and reconstruction kernel on emphysema quantification: relationship to the magnitude of the CT emphysema index. *Acad Radiol.* 2010;17(2):146-156.

44. You C, Yang L, Zhang Y, Wang G. Low-dose CT via deep CNN with skip connection and network-in-network. International Society for Optics and Photonics; 2019:111131W.

45. You C, Cong W, Vannier MW, et al. CT super-resolution GAN constrained by the identical, residual, and cycle learning ensemble (GAN-CIRCLE). *IEEE Trans Med Imaging.* 2019;39(1):188-203.

46. You C, Cong W, Wang GE, et al. Structurally-sensitive multi-scale deep neural network for low-dose CT denoising. *IEEE Access.* 2018;6:41839-41855.

47. Mansoor A, Perez G, Nino G, Linguraru MG. Automatic tissue characterization of air trapping in chest radiographs using deep neural networks. *Conf Proc IEEE Eng Med Biol Soc.* 2016;08(2016):97-100. https://doi.org/10.1109/EMBC.2016.7590649.

48. Efron B. Bootstrap methods: another look at the jackknife. Breakthroughs in statistics. Springer. 1992;569-593.

SUPPORTING INFORMATION

Additional supporting information may be found online in the Supporting Information section.

How to cite this article: Mühlberg A, Kärgel R, Katzmann A, et al. Unraveling the interplay of image formation, data representation and learning in CT-based COPD phenotyping automation: The need for a meta-strategy. *Med Phys.* 2021;48:5179–5191. https://doi.org/10.1002/mp.15049

SCIENTIFIC
REPORTS
natureresearch

There are amendments to this paper

OPEN

The Technome - A Predictive Internal Calibration Approach for Quantitative Imaging Biomarker Research

Alexander Mühlberg[1*], Alexander Katzmann[1,6], Volker Heinemann[3,4], Rainer Kärgel[1], Michael Wels[1], Oliver Taubmann[1], Félix Lades[1], Thomas Huber[2], Stefan Maurus[2], Julian Holch[3,4], Jean-Baptiste Faivre[5], Michael Sühling[1], Dominik Nörenberg[2,7] & Martine Rémy-Jardin[5,7]

The goal of radiomics is to convert medical images into a minable data space by extraction of quantitative imaging features for clinically relevant analyses, e.g. survival time prediction of a patient. One problem of radiomics from computed tomography is the impact of technical variation such as reconstruction kernel variation within a study. Additionally, what is often neglected is the impact of inter-patient technical variation, resulting from patient characteristics, even when scan and reconstruction parameters are constant. In our approach, measurements within 3D regions-of-interests (ROI) are calibrated by further ROIs such as air, adipose tissue, liver, etc. that are used as control regions (CR). Our goal is to derive general rules for an automated internal calibration that enhance prediction, based on the analysed features and a set of CRs. We define qualification criteria motivated by status-quo radiomics stability analysis techniques to only collect information from the CRs which is relevant given a respective task. These criteria are used in an optimisation to automatically derive a suitable internal calibration for prediction tasks based on the CRs. Our calibration enhanced the performance for centrilobular emphysema prediction in a COPD study and prediction of patients' one-year-survival in an oncological study.

Technical variation poses a problem for radiological quantification of biological structures – in particular in terms of morphological tissue characteristics. In computed tomography (CT), for instance, a change in scan protocol or reconstruction method may considerably vary observable texture in the acquired 3D image series and thus texture quantifying features. Therefore, status-quo radiology is mainly a qualitative effort in the sense that it relies on the radiologists' experience who can usually integrate above-mentioned technical variation intuitively in their diagnosis. Correspondingly, most radiological tomographic reconstruction methods, also in magnetic resonance imaging (MRI), are optimised for qualitative, not quantitative, assessment. For the successful application of radiomics, high-throughput and high-content screening of standard-of-care medical images, or quantitative imaging biomarkers (QIB) research in general, though, this technical variation poses a much larger problem. Analysed features can be very sensitive to the impact of technical variation. That is, a feature can be strongly affected by a technical characteristic, e.g., by a slight streak artifact barely recognisable for the human reader. It is hence worth striving for appropriate techniques to reduce this negative impact of technical variation on extracted features and, moreover, to make subsequent statistical analysis resistant to such effects. In the scientific literature, mainly the impact of trivially reducible technical variation on features is analysed, e.g., by the Quantitative Imaging

[1]Department CT R&D Image Analytics, Siemens Healthineers, Forchheim, 91301, Germany. [2]Department of Radiology, University Hospital Großhadern, LMU, Munich, 81377, Germany. [3]Department of Medical Oncology, University Hospital Großhadern, LMU, Munich, 81377, Germany. [4]Comprehensive Cancer Center, University Hospital Großhadern, LMU, Munich, 81377, Germany. [5]Department of Thoracic Imaging, CHRU et Universite de Lille 2, Hospital Calmette, Lille, 59037, France. [6]Neuroinformatics and Cognitive Robotics Lab, University of Technology, Ilmenau, 98693, Germany. [7]These authors jointly supervised this work: Dominik Nörenberg and Martine Rémy-Jardin. *email: alexander-muehlberg@hotmail.com

Biomarker Alliance (QIBA)[1]. In CT, this variation results from varying scan and reconstruction parameters or from acquisitions with entirely different scanners. For the most part, it is thus reducible by simply setting the associated parameters to constant values. Accordingly, the QIBA's main goal is the optimisation and standardisation of scan protocols. This is often attempted by impact analysis of acquisition parameters or scanner types[2–7].

Technical variation can however result from interaction between the image acquisition and individual patient characteristics, too, yielding both inter-patient noise and artifact variation. In CT, a corpulent patient with a larger cross-section will usually have a higher noise level within his body than a slimmer patient. This effect appears despite constant scan parameters as less quanta arrive at the detector. Another example is beam hardening that is stronger if the cross-section of the patient is larger: photons in the center then have a higher average energy than in the periphery. As this kind of technical variation can occur although extrinsic factors are kept constant, we call it the non-(trivially-)reducible technical variation. While regular technical variation results from variation of e.g. voxel spacing, reconstruction kernel or slice thickness, non-reducible technical variation is a result of variation of patient geometry and/or attenuation characteristics and expresses in noise or artifact variation, such as cupping artifacts. While scientists in the field of image acquisition and image reconstruction deal with such non-reducible technical variation decreasing its qualitative influence with advanced techniques (e.g. via tube current modulation[8]), it is not known which impact non-reducible technical variation has on derived quantitative features, e.g. radiomics, in relation to the examined biological or pathological variation.

The impact of technical variation on the evaluation of scientific questions was initially marked by Leek[9]. He has shown that for laboratory experiments the measurements were correlated with e.g. the date of the experiment. He therefore used the word surrogate as a feature indicative for the technical variation as the date or the humidity of the laboratory. He describes different data-driven procedures to identify such surrogates. Fortin expanded this concept and used the cerebrospinal fluid in MR images as a so-called control region (CR, for technical variation)[10]. A singular value decomposition (SVD) of the CR cohort variation is used to determine the main technical variation in the cohort. He then decomposed the voxel intensity distribution of the brain into an impact of the biological label and an impact of the main technical variation by a least-squares fit. Finally, the intensity distribution is adjusted for the fitted technical variation, which is why the method is called Removal of of Artificial Voxel Effect by Linear Regression (RAVEL). ComBAT[11] is also an older technology from genomics that found attention recently as the method was capable of stabilising radiomics features for technical variation resulting from different imaging protocols[12]. In this method, a feature is decomposed into an additive and an multiplicative imaging protocol effect. The effects are then estimated by an empirical Bayes fit and removed from the feature by subtraction and division. These are statistical methods, i.e. a calibration is learned on the same data as it is applied and their main focus is the stabilisation of intensity distributions or features with regards to technical variation.

Surrogate features encoding technical variation also play a crucial role in our approach. Although technical variation can completely falsify a statistical analysis, in-house experiments show that machine learning classifiers such as a random forest can to a certain degree automatically learn and therefore compensate for technical variation when predicting a label if enough data and features are available. Our goal is therefore to automatically qualify and subsequently select surrogates from CRs to enhance prediction tasks associated with the actual target regions. We focus on predictive calibration with regards to a label in contrast to the statistical standardisation found in the literature. Whereas statistical standardisation learns the calibration on the same data as it is applied, we apply the calibration to unseen data.

Method

As mentioned above we assume that the impact of both types of considered technical variation is not only present in a target region-of-interest (ROI), but also in a CR, a region inside the body which should ideally show only little inter-patient biological variation. The CR is thus assumed to store a patient-specific fingerprint of the inter-patient technical variation. Ideally, a CR should be close to a ROI in order to reduce the influence of spatially non-uniform noise and artifact distributions. Chambers of in-scan phantoms, which often serve as CRs, additionally placed next to the patient at scan time are particularly subject to this kind of inhomogeneity[13]. Besides, most clinical data is not acquired with in-scan phantoms. Our approach therefore merely relies on CRs that are naturally part of the imaging data to be processed. Regarding patient cohorts, the fingerprints of both regions, ROI and CR, induce an inter-patient correlation of biological and technical image information. For the sake of robustness, we extract surrogates from a multitude of CRs. We assume the entirety of surrogates over all CRs to contain the essential reducible and non-reducible technical information for relational quantification such as shown in Fig. 1. However, not all CRs and especially not all surrogates extracted from CRs are indeed suitable to represent technical variation. Thus they need to be qualified for this purpose, which we will describe later.

Reducible technical variation is termed T^R and non-(trivially-)reducible technical variation T^{NR}. We call the entirety of features in a ROI a radiome $(r_{ij})_{j\in R} := (r_{ij}, .., r_{iM})^\top \in \mathbb{R}^M$ with patient index i and feature index j for a total of M extracted features. The relevant clinical annotation (or label) for patient i is termed b_i. The general task of personalised medicine within imaging science is the generation of classification models which predict (machine learning) or explain (statistics) b_i by analysing $(r_{ij})_{i\in P, j\in R}$ for the analysed patients P. This allows to design decision-support algorithms $\hat{b}_i = f^{\mathrm{clf}}((r_{ij})_{j\in R})$, where R is the set of used features. We define S as the set of used surrogates. Accordingly the surrogate space is termed s_{il} for $l \in S$.

Explicit and implicit calibration in literature: *stabilisation* and *predictive* mode.
To understand the dichotomy we will introduce in our calibration, we first have to explain the case separation of explicit and implicit calibration in the literature. It is a well-known problem that a calibration method that maximises the stability against technical variation may not be the method that enables optimal classification or prediction[14–16]. This is based on the fact that the focus of a calibration for the latter must lie on the discriminative part of the biological variation – and not the overall stability. We thus see two different operational modes of a calibration: *predictive*

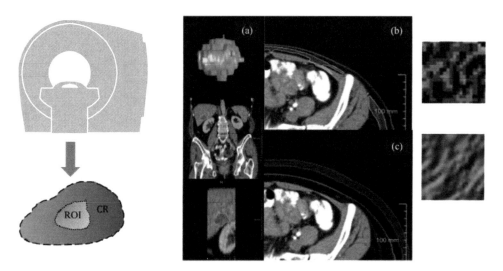

Figure 1. A tumour ROI (**a**) is internally calibrated by relating it to texture measurements in CRs such as adipose tissue (**b**) or air (**c**). It is assumed that T^R and T^{NR} are present in ROI and CRs.

mode and *stabilisation* mode. The *stabilisation* mode yields the stabilised radiome only, while the *predictive* mode works with respect to the final prediction task.

The *stabilisation* mode is used to maximise the amount of information invariant to technical variation. RAVEL and ComBat use this mode as they explicitly remove the technical variation from the features. They first decompose the feature into technical and biological covariates. This happens by a linear fit. RAVEL fits the intensity distribution by the singular vectors of the CR intensities as determined by a SVD. The decomposition of RAVEL for the voxel $\vec{v}$ is,

$$I_i(\vec{v}) = \beta_0 + \beta_b b_i + \sum_k^N \beta_k \mathrm{SVD}_k(I_i(CR(\vec{v}))). \tag{1}$$

This fit yields the estimates $\hat{\beta}_0$, $\hat{\beta}_b$ and $\hat{\beta}_k$, $k = 1, \dots N$. The standardised intensity distribution then is,

$$I_i^*(\vec{v}) = I_i(\vec{v}) - \left(\hat{\beta}_k + \sum_k^N \hat{\beta}_k \mathrm{SVD}_k(I_i(CR(\vec{v}))) \right). \tag{2}$$

Translating the RAVEL principle from intensity distributions to features, we arrive at,

$$r_{ij}^* = r_{ij} - \left(\hat{\beta}_0 + \sum_k^N \hat{\beta}_k \mathrm{SVD}_k((s_{il})_{i \in P, l \in S}) \right). \tag{3}$$

We term this approach RAVEL-like. Alternatively, the impact of covariates can linearly be removed after decorrelating them to select the most important ones: With ANCOVA, the feature can be decomposed into covariates[17]. Then a GLM fits the deviation of the feature from a fixed value by the deviation of covariates from their the average value within the cohort. By using surrogates as covariates this becomes $\Delta s_{il} = s_{il} - \mathbf{avg}_i((s_{il})_{i \in P})$. In analogy to this covariate adjustment, the impact of qualified surrogates on the feature can be estimated by a linear fit,

$$r_{ij}^* = r_{ij} - \left(\hat{\beta}_{0j} + \sum_{l \in S} \hat{\beta}_{lj} \Delta s_{il} \right) \tag{4}$$

for surrogates s_{il}, $l \in S$ and with the calibrated feature value r_{ij}^* for $j \in R$. We see that the general form of the explicit calibration is,

$$r_{ij}^* = r_{ij} - g_j^{\mathrm{reg}}. \tag{5}$$

In RAVEL and ANCOVA, the function g_j^{reg} takes a linear form and is parametrised by singular vectors or covariates. The regression coefficients are determined by a decomposition of the feature and a linear fit. This approach however is per design extensible for a machine learning-based calibration procedure. The feature can of course also be fitted, or trained, by a machine learning model. For instance, g_j^{reg} could be a random forest regression g_j^{RF} that predicts the deviation of the feature from a fixed value by the deviation of the surrogates from their mean. Subsequently, the predicted value can be subtracted from the feature value of the test data. This approach

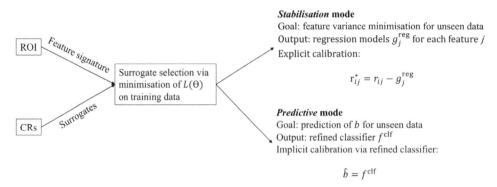

Figure 2. Overview of the method.

can be seen as a machine learning generalisation of the linear covariate adjustment. We will need this analogy later on for the validation of our method.

The *predictive* mode, on the other hand, assumes that not the whole image information needs to be stabilised for technical variation, but solely the image information that is needed for the classification task. Thus the invariance is optimised for the diagnostic relevant information. This approach is motivated by Leek[9], who suggests to incorporate suitable surrogates for technical variation in the classification process. He also points out that it is important to identify suitable surrogates. In the field of laboratory experiments, surrogates such as 'date of experiment' or 'laboratory personnel' can be identified by a data-driven analysis. We, however, want to search CRs for suitable surrogates. We term the incorporation of these surrogates the *predictive* calibration mode. The classification process to predict the label b_i for patient i then becomes,

$$\hat{b}_i = f^{\mathrm{clf}}\left((r_{ij})_{j \in R}, (s_{il})_{l \in S}\right), \tag{6}$$

where $(s_{il})_{l \in S}$ are surrogate values for the patient i. The radiome is implicitly calibrated within the classification process by incorporating the qualified surrogates that have been shown to interact with the radiome technically and linearly.

Naturally, not all available surrogates should be used for explicit or implicit calibration, but only a qualified subset that is indeed suitable to calibrate the given features. Using all accessible surrogates without any qualification is termed a naive approach. Given enough data a classifier can to some degree compensate for technical variation even in a naive approach, however, we expect an increase in stabilisation and especially predictive performance when the surrogates are qualified for the given features which they shall explicitly or implicitly calibrate.

We term the refined classifier $f^{\mathrm{clf}}\left((r_{ij})_{j \in R}, (s_{il})_{l \in S^q}\right)$ based on features and qualified surrogates S^q 'technome *predictive* mode'. The set of regression models g_j^{reg}, $j \in R$ to explicitly calibrate a feature based on qualified surrogates for the respective feature is termed 'technome *stabilisation* mode'. The technome can therefore have two modes, specialising on stabilisation or prediction. We now describe how the surrogates are qualified and the technome is constructed based on surrogate qualification for the respective mode.

Technome Construction as a Model-Based Optimisation

A diagram that gives an overview of this section is shown in Fig. 2. As explained above, used surrogates should not only be qualified for a suitable calibration but also enhance prediction. We denote a qualification score of a surrogate $l \in S$ for the feature $j \in R$ as q_{jl}. Surrogates are qualified by *in vivo* assessment q_{jl}^{inVivo}, *in silico* tests $q_{jl}^{\mathrm{inSilico}}$, phantom tests $q_{jl}^{\mathrm{inVitro}}$ and also for statistical reasons q_{jl}^{orthog}. An exhaustive explanation and rationale of our qualification criteria can be found in the Supplementary Material (A.). Pseudo code of the method is found in Supplementary Material (B.). As the relevance of the different surrogate qualification criteria for the technome construction process is unknown, we have to introduce free weighting parameters $\Theta = \{\theta^{\mathrm{inVivo}}, \theta^{\mathrm{inSilico}}, \theta^{\mathrm{inVitro}}, \theta^{\mathrm{orthog}}\}$ which we determine on training data. We therefore define a loss function that has to be minimised,

$$L(\Theta) = L_{\mathrm{train}}(\Theta) + L_{\mathrm{calib}}(\Theta). \tag{7}$$

The performance for the respective task on the training data is introduced as L_{train}. In *predictive* mode it should be a classification performance metric e.g. an Area-Under-Curve of the receiver-operating characteristic (AUC) for the given label, in *stabilisation* mode it should ideally be a metric that quantifies the stabilisation performance. For the technome *stabilisation* mode, the training loss may be introduced as the variance of the feature deviation from the cohort mean that can be explained by the surrogates' deviation of the cohort mean. For the technome *predictive* mode, the training loss may be the predictive performance in a 10-fold cross-validation (CV) on the training data in case of a machine learning procedure. In case of a statistical classifier such as the logistic regression, we can simply use the discriminative AUC of the classifier on the training data.

We now calculate a qualification score Q_{jl} for each of the surrogates $l \in S$ for each of the features $j \in R$ of the radiome. We define the qualification score for surrogate l for the feature j and the parametrisation Θ as,

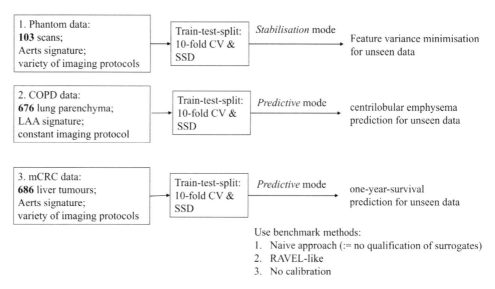

Figure 3. Overview of the validation strategy.

$$Q_{jl}(\Theta) = \theta^{\text{inVivo}} q_{jl}^{\text{inVivo}} + \theta^{\text{inSilico}} q_{jl}^{\text{inSilico}} + \theta^{\text{inVitro}} q_{jl}^{\text{inVitro}} + \theta^{\text{orthog}} q_{jl}^{\text{orthog}}, j \in R, l \in S. \tag{8}$$

The set of qualified surrogates to calibrate the respective feature of the radiome is then defined as all surrogates with a qualification score higher than an arbitrary value of $Q_{min} = 1.0$. The weights Θ control which surrogates are selected for the internal calibration.

$$S_{R,\Theta}^q = \{l \in S | Q_{jl}(\Theta) > Q_{min}, j \in R\}. \tag{9}$$

For each feature $j \in R$ of the radiome a subset of qualified surrogates is determined. By penalising large weights Θ, only surrogates with sufficiently high qualification values are selected by Eq. 9, thus semantically constraining the available image information used for calibration. The crucial point is now to enforce an internal calibration of the features by introducing the calibration loss as,

$$L_{\text{calib}}(\Theta) = -\alpha / \left(\sum_{\theta \in \Theta} \theta \right), \tag{10}$$

in which a higher qualification of the surrogates translates to a lower value of L_{calib}. The design parameter α is introduced to find a reasonable tradeoff between the used performance metric on the training data and the enforced internal calibration. A Bayesian optimisation process[18] is used to determine the weights of the semantic constraints Θ that minimise the combined loss for training and qualification $L = L_{\text{train}} + L_{\text{calib}}$ on the experimental data similar to systems-biology model building approaches[19],

$$\Theta = \underset{\Theta}{\text{argmin}}(L_{\text{train}} + L_{\text{calib}}). \tag{11}$$

The minimisation of L is achieved by Bayesian optimisation on training data. This determines the chosen weights Θ which in turn construct the technome.

We expect a higher predictive and stabilisation performance for unseen data when the surrogates have higher qualification and indeed internally calibrate the analysed radiome.

As a side effect, the technome can also be read out to discover new insights of feature stability in general (*stabilisation* mode) or technical impact on diagnosis (*predictive* mode). The qualification criteria scores Q_{jl} on the one hand help to determine suitable surrogates, on the other hand they also determine feature stability *in vivo*, *in vitro* and *in silico*. If many qualified surrogates can be found for a feature even for low values of Θ, the feature is non-robust. Finally, the weights Θ can help to explore whether either phantoms, simulations or associations in real data are most important to identify a suitable internal calibration that enhances predictive performance. The importance of different qualification criteria is to the best of our knowledge not yet analysed.

Validation Strategy

A diagram that gives an overview for this section is shown in Fig. 3. We employ a validation scheme similar to that of RAVEL[10] that assesses if a calibration increases reproducibility of already shown associations between features and biological labels. For this purpose, we use two signatures that were often shown to be predictive for their use case: the Aerts signature[20] for patient survival prediction in oncology and the low-attenuation area (LAA) signature consisiting of LAA910 and LAA950 for emphysema assessment[21]. For the Bayesian optimisation needed to

Study type	Task	Data	T^R	T^{NR}	Surrogates	Features	Technome Mode
Phantom	Feature Stabilisation	103 CT Scan protocols for phantom	X	X	PyRadiomics features in CRs air, adipose tissue, liver, trachea, spleen, heart, aorta	Aerts signature in phantom 'liver-tumours'	*Stabilisation*
COPD	CLE Prediction	676 lung parenchyma on CT		X		LAA signature in parenchyma	*Predictive*
	CLE SSD			X			*Predictive*
mCRC	1-ys Prediction	686 liver tumours on CT	X	X		Aerts signature in liver tumours	*Predictive*
			X	X			*Predictive*
	1-ys SSD		X	X			*Predictive*

Table 1. Overview of all experiments. Shown are the clinical field, task, used data, technical variation within the data, i.e. non-reducible T^{NR} or also reducible T^R, used surrogates, features and technome mode. CLE: centrilobular emphysema, 1-ys: one-year-survival.

Validation Scheme	Technome Computation and Training Set	Test Set
10-fold CV	9/10	1/10
SSD	1/5	4/5

Table 2. Validation schemes and the ratio of used data for training and test set.

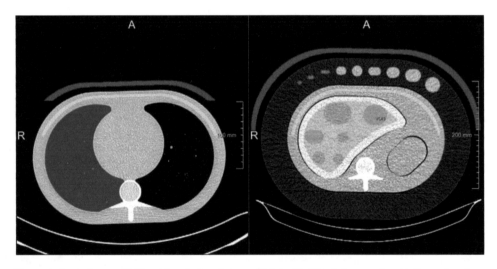

Figure 4. Lung phantom (left) and liver-lesion phantom (right) with CRs air (magenta) and 'adipose tissue' (yellow).

construct the technome, we integrated an established Python implementation[22] in our pipeline. The acquisition function 'upper confidence bound', 8 initial seeds, 500 iterations and a Kappa of 5 as tradeoff between 'exploration' and 'exploitation' were chosen. An overview of all conducted experiments for *stabilisation* and *predictive* mode is shown in Table 1. Analyses were conducted with R packages (version 3.3.2, www.R-project.org) and scikit-learn[23].

Benchmark methods and train-test split. We compare the performance in technome *stabilisation* and *predictive* mode with the RAVEL-like approach and the naive approach, i.e. using all surrogates without any qualification criteria, as introduced above.

As a first test we compare the technome performance with a naive approach. As explained above a naive approach uses Eq. 5 or 6 respectively with all available surrogates without constrainining them. It is assessed whether the classifier or regressor can identify the relevant surrogates by themselves without additional qualification criteria. We test two cases for the stabilisation and prediction: a simple linear model f^{GLM} or g^{GLM} and a more complex, non-linear random forest classification, f^{RF}, or regression, g^{RF}, model. To minimise redundant surrogates for the naive linear approach, surrogates are decorrelated via minimum redundancy maximum relevance (mRMR) algorithm[24,25] and finally regularised by best subset feature selection according to the Akaike information criterion (AIC) to yield the set of used surrogates S.

The RAVEL-like approach uses the principal components of all available surrogates and uses formula 3 for the explicit stabilisation. For the predictive mode, the principal components are incorporated in the classifier. The

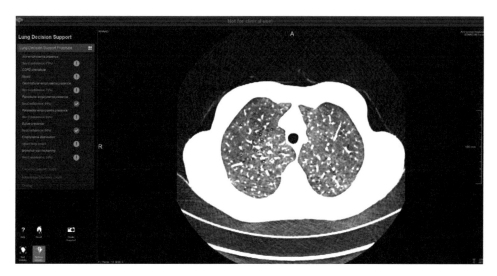

Figure 5. Prototype used for parenchyma analysis. An example of a COPD patient is shown.

principal components of the CR's surrogates are determined via SVD. Principal components are collected until they explain 95% of the variance of the training data.

For technome *predictive* or *stabilisation* mode, we only assess linear approaches f^{GLM} and g^{GLM} respectively. By concept, we explicitly enforce a linear calibration with the inVivo and inSilico qualification and expect a potential advantage to use a regularised, interpretable and statistically valid model on training data.

To test the stabilisation or prediction performance on unseen data, a 10-fold CV is used. Therefore, the data is split in training and test data, where the ratio of train to test data is 9:1. The training data is used to qualify the surrogates and optimise the weights Θ. This yields the technome *stabilisation* or *predictive* mode, i.e. the regression model for explicit feature calibration, g^{GLM}, or the refined classifier for the medical label, f^{GLM}. The models are then applied for the unseen test data and the predictive or stabilisation performance is assessed. As very often in clinical scenarios only a small number of patients can be collected to detect the effect of a drug etc., the performance of the method in the small sample setting is assessed. We therefore choose the ratio of train to test data as 1:4 and term this scenario small sample detectability setting (SSD) (Table 2). Both train-test split scenarios are summarised in Table 2.

Phantom dataset - feature stabilisation. To validate the technome *stabilisation* mode, 103 CT images of a liver-lesion phantom[26] acquired with a Siemens Somatom Zoom Scanner are analysed (Fig. 4, right). The images were acquired with a variety of scan and reconstruction parameters. Varied parameters are e.g. kvp, mAs, slice thickness, voxel size. The 3 texture features of the Aerts signature are extracted within a 'tumour-like' ROI and the 844 PyRadiomics[27] features are used as surrogates S for the CRs air and 'adipose tissue'. The ROIs are segmented by a semi-automated algorithm[28].

The training loss is defined as variance R^2 of feature j that is explained by g^{GLM} for the current qualified surrogate selection averaged over all analysed features $L_{train} = -\mathbf{avg}_j((R^2(g_j^{GLM}))$. For the calibration loss, only the qualification criterion q_{jl}^{inVivo} is used for phantom scans, as no biological induced variation needs to be excluded. Therefore $L_{calib} = -\alpha\theta^{inVivo}$ with $\alpha = 0.1$ is enforced. The labels of the scan and reconstruction parameters are not used for the task.

The stabilisation performance on the test data is assessed by the variance reduction of the procedure i.e. the difference of feature variance before and after calibration $\text{Perf}_j^{stabil} = 1 - \frac{\text{var}((r_{ij}^*)_{i\in p})}{\text{var}((r_{ij})_{i\in p})}$. The performance is then determined as the average performance for all features $j \in R$, $\text{Perf}^{stabil} = \mathbf{avg}_j(\text{Perf}_j^{stabil})$.

COPD dataset – centrilobular emphysema prediction. To validate the technome *predictive* mode ability to enhance centrilobular emphysema prediction in the presence of only non-reducible technical variation, 676 partly contrast-enhanced CT scans of 676 patients (age: 84.1 ± 14.8 y) each reconstructed with a soft (B36) and hard (B71) kernel, i.e. 1352 clinical CT images were acquired with a Siemens Somatom Force in Lille, France[29]. The cohort consists of patients with different symptoms of chronic obstructive lung disease (COPD). The study was approved by the local ethics commitee with waiver of the informed consent because CT examinations were part of routine clinical practice. Scan and reconstruction parameters are kept constant and only images reconstructed with kernel B71 are analysed, therefore no reducible technical variation exists. Additionally, tube current modulation techniques[8] CareDose and CareKV were used to guarantee consistent image quality and further minimise non-reducible technical variation.

The CRs air, trachea, adipose tissue, liver, heart, spleen and aorta segmented by an automated algorithm by Seifert *et al.*[30]. The lung parenchyma is segmented by an adaption of the deep learning segmentation of Yang *et al.*[31] (Fig. 5). As features R we analyse the LAA signature. The LAA signature consists of two clinical standard features LAA910 and LAA950[21]. These features are commonly quantified to assess emphysema in clinical practice. Surrogates S are the 844 PyRadiomics[27] features extracted in the CRs.

For the inSilico surrogate qualification q_{jl}^{inSilico}, *in silico* variation was generated by superimposition of Gaussian, Rayleigh, Poisson and Gamma noise with 5 steps of monotonically increased noise each. For the inVitro surrogate qualification q_l^{inVitro}, 16 CT images with varying scan and reconstruction parameters of the parenchyma in an anthropomorphic lung phantom[32] (Fig. 4, left) were used.

The training loss is defined by the discriminative AUC of f^{GLM} for the current selection of qualified surrogates on the training data $L_{\text{Train}} = -\text{AUC}$. The calibration loss is defined as $L_{\text{calib}} = -\alpha/(\sum_{\theta \in \Theta} \theta)$ with all qualification criteria and $\alpha = 0.2$. The predictive performance is assessed as the AUC on the test data.

We compare the predictive performance of the technome for the hard kernel with the predictive performance of the LAA signature on the soft kernel. While a hard kernel is clinically considered inadequate for emphysema assessment due to larger inter-patient noise variation, we expect a predictive performance similar to the results for a soft kernel after calibration.

mCRC dataset – one-year-survival prediction.

To validate the technome *predictive* mode ability to enhance one-year-survival prediction in the presence of non-reducible and also reducible technical variation, 118 contrast-enhanced CT scans of 75 patients (age: 61.9 ± 11.4 y) with 686 analysed liver tumours acquired with a variety of scanners from different vendours (GE, Philips, Siemens, Toshiba) in Munich, Germany, are analysed[33]. The cohort consists of metastatic colorectal cancer (mCRC) patients with liver metastases. The study was approved by the local ethics commitee with waiver of the informed consent because CT examinations were part of routine clinical practice. Scan and reconstruction parameters show large variation i.e. reducible and non-reducible technical variation is present. Tumours were segmented by a semi-automated segmentation algorithm[28]. The Aerts signature[20] consisting of 4 features R is extracted from the tumour ROIs. Feature extraction and CR segmentation was conducted within a specialised radiomics framework[34] (Fig. 6). Used surrogates S are the PyRadiomics features[27] in the CRs air, trachea, adipose tissue, liver, heart, spleen and aorta again segmented by the algorithm of Seifert *et al.*[30].

In silico variation for the calculation of q_{jl}^{inSilico} was generated by superimposition of Gaussian, Rayleigh, Poisson and Gamma noise with 5 increasing steps each. For *in vitro* variation, 103 CT images of an anthropomorphic liver-lesion phantom[26] (Fig. 4, right) with a variety of scan and reconstruction parameters were used.

Training and test loss are defined in analogy to the COPD dataset. For each fold, the tumours of a patient were either in the training or the test data and a split was not allowed. Again, the predictive performance is assessed as the AUC on the test data.

For comparison purposes, all PyRadiomics features are extracted in the tumours and entered in a random forest classifier to predict one-year-survival. This random forest is optimised by hyperparameter tuning on the training data. With this step, we can assess, whether our calibration outperforms a hyperparameter tuned classifier using an abundance of features within the ROI. Finally, we compare the predictive performance of our technome classifer with an advanced deep learning method. For more details regarding the deep learning architecture we refer to Katzmann *et al.*[35].

Technome discovery – understand importance of qualification criteria and surrogates.

Finally, it has to be analysed how qualification criteria are associated with the predictive performance. It is not clear whether an improvement of the predictive performance is really induced by an enforced internal calibration or simply by additional accessible biological information from the CRs. Therefore we have to assess whether qualification criteria indeed enforce an internal calibration. If this is the case, a higher qualification of surrogates should translate to a higher predictive performance. For the data presented above, each qualification criterion is inspected individually. For a first assessment, no loss is minimised as the goal is to analyse the association of qualification criteria with predictive performance without emphasising a certain subspace of the weight space Θ. For the inspection of individual qualification critera, Q_{min} is varied to generate 100 samples starting with 4 random seeds. Also combinations of qualification criteria are assessed for their association with predictive performance.

In a second step, it is assessed whether the minimisation of the loss L with $\alpha = 0.1$ and only one qualification criterion, e.g. $L_{\text{calib}} = -0.1/\theta^{\text{inVivo}}$, yields acceptable predictive performance for each qualification criterion individually.

Finally, we read out the surrogates for calibration that yield the best performance for prediction of centrilobular emphysema and one-year-survival. Potentially, the integration of these surrogates could enhance clinical models using the LAA or Aerts signature.

Results

Phantom dataset – feature stabilisation.

The results are shown in Fig. 7. The technome *stabilisation* mode gives a stabilisation performance [% variance reduction on test set] in a 10-fold CV (SSD) of 90.4% (91.8%). The RAVEL-like calibration yields 79.3% (76.7%) and the naive approach via GLM 74.7% (79.3%) or via random forest 76.7% (42.7%).

Apparently, 21 training datasets of the SSD setting were not sufficient for a complex calibration as observed by the stabilisation performance of the random forest regression with all surrogates.

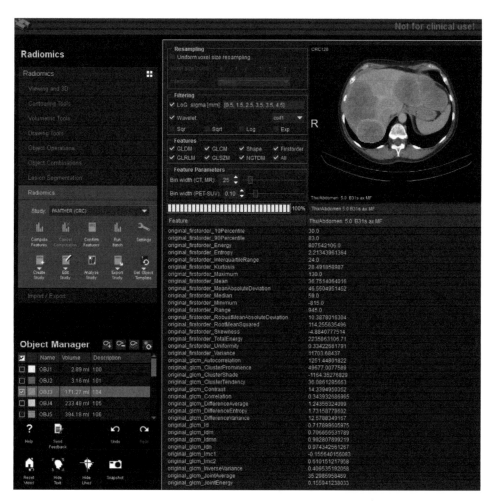

Figure 6. Radiomics prototype[34] for PyRadiomics feature and surrogate extraction. An example of a mCRC patient is shown.

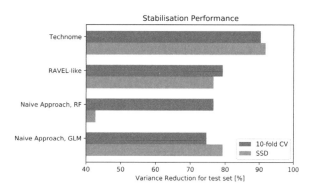

Figure 7. Stabilisation performance [% variance reduction] in 10-fold CV (blue) and SSD (red) in the phantom study.

COPD dataset – centrilobular emphysema prediction. The results are shown in Fig. 8. The predictive performance [% AUC on test set] in 10-fold CV (SSD) improved to 81.5% (74.4%) with technome *predictive* mode in comparison to GLM with 60.7% (60.0%) or random forest with 58.1% (59.1%). The technome was also superior to a RAVEL-like calibration of 70.4% (64.6%) and the naive approach via GLM with AUC of 70.8% (65.4%) or via random forest with 74.8% (67.0%).

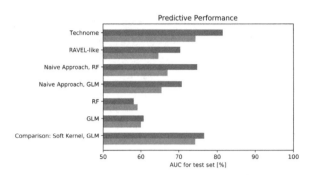

Figure 8. Predictive performance [% AUC] in 10-fold CV (blue) and SSD (red) for centrilobular emphysema prediction in the COPD study.

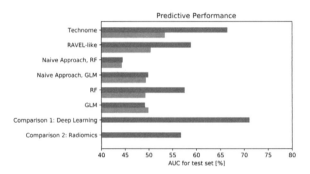

Figure 9. Predictive performance [% AUC] in 10-fold CV (blue) and SSD (red) for one-year-survival prediction in the mCRC study.

As a comparison, the predictive performance of LAA signature on the soft kernel B36, which is the clinical standard for emphysema classification, was 76.6% (74.3%), which is numerically inferior to the technome calibration for the hard kernel B71.

mCRC dataset – one-year-survival prediction. The results are shown in Fig. 9. The predictive performance [% AUC on test set] in 10-fold CV (SSD) improved to 66.5% (53.4%) with technome *predictive* mode in comparison to GLM with 49.2% (49.9%) or random forest with 57.6% (49.3%). The technome was also superior to a RAVEL-like calibration of 58.9% (50.4%) and the naive approach via GLM with AUC of 49.9% (49.2%) or random forest with 44.5% (43.3%). In the SSD scenario, no approach yielded an acceptable predictive performance. Apparently, the variation is too large to calibrate the signature with only few data.

For comparison, an advanced deep learning approach based on sparse autoencoder pre-training[35] optimised for the same data achieved a performance of 71.1%. A radiomics approach with a random forest highly optimised by hyperparameter tuning using a nested CV and all PyRadiomics features calculated within the tumour ROIs achieves 56.8%.

Technome discovery – understand importance of qualification criteria and surrogates. In a first step, the association of predictive performance with each qualification criterion individually was inspected. Apparently, only inSilico and inVivo qualification show a deterministic behaviour for high qualifications. This is shown exemplarily for centrilobular emphysema and the inSilico qualification in Fig. 10 on the left. The inVitro and orthogonality qualification used alone, however, are not associated with predictive performance. This is shown exemplarily for the inVitro qualification on the right in Fig. 10. When the loss L is minimised, the tradeoff between training loss and calibration loss used in the Bayesian optimisation detects a point of good predictive performance for the inVivo and inSilico qualification within the deterministic area ('Bayes' in Fig. 10).

Therefore, we inspect the association of combined inVivo and inSilico qualification weights with predictive performance in Fig. 11. For the inspection of combined inVivo and inSilico weights, $\Theta = \{\theta^{\text{inVivo}}, \theta^{\text{inSilico}}\}$, 500 random samples are drawn uniformly with fixed $Q_{\text{min}} = 1.0$. As can be seen in Fig. 11 the predictive performance is lower when all or many surrogates are used (underconstrained, UC) or no surrogate is qualified enough and thus no calibration is used (overconstrained, OC). Please note that surrogates with the highest qualification are found in direct border to the OC region and lower weights translate to a higher qualification of the surrogates as a consequence of Eq. 9. When only non-reducible technical variation is present, i.e. scan and reconstruction

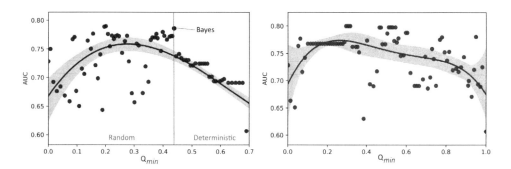

Figure 10. Association of the inSilico (left) and inVitro qualification (right) with the predictive performance for centrilobular emphysema. For high qualification the inSilico qualification criterion leads to a higher correlation of predictive quality with qualification, while the correlation of the predictive performance rapidly decreases for lower qualification. The inVitro qualification criterion when used alone, however, shows no obvious association with predictive performance. Only for inVivo and inSilico qualification criteria, the Bayesian optimisation constructed a technome calibration with a good predictive performance ('Bayes').

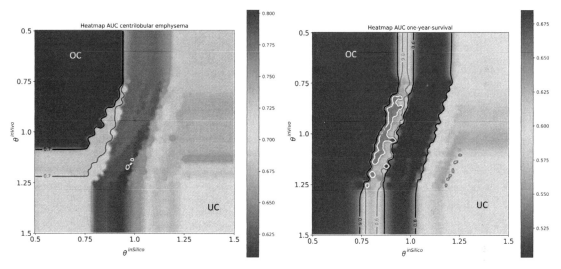

Figure 11. Association of combined inVivo & inSilico qualification weights with the predictive performance [%AUC on test data] for centrilobular emphysema (left) and one-year-survival (right). Overconstrained (OC), i.e. no calibration, and underconstrained (UC) regions, i.e. also surrogates with low qualification are used, show lower predictive performance than models, where the analysed features were implicitly calibrated by highly qualified surrogates. Lower weights Θ translate to a higher qualification.

parameters are constant within the study, the inSilico qualification appears to be more important for the predictive performance as observed for centrilobular emphysema prediction in comparison to one-year-survival prediction, where also reducible technical variation is present. A good predictive performance for one-year-survival was achieved when inSilico and inVivo qualification were combined. A high qualification according to both criteria translates to a high predictive performance, while a larger weight for the inSilico qualification guarantees a higher predictive performance for emphysema presence.

In both studies a high qualification of surrogates according to the defined qualification criteria is associated with enhanced predictive performance. Though their individual inspection did not show any association with predictive performance, the additional integration of inVitro and orthogonality qualification still increased predictive performance numerically in comparison to the combined inVivo and inSilico qualification.

The surrogates that guaranteed the highest AUC on test data are shown in Table 3. The LAA signature is calibrated best by texture metrics within a cylindrical CR in the trachea, while the Aerts signature is calibrated by a combination of texture metrics in liver and heart. This could potentially be due to applied contrast injection variation that has to be calibrated.

Signature	CR	Qualified surrogates
Aerts signature		
Grey Level NonUniformity	Heart	WL_LHH_GLSZM_LargeAreaEmphasis
	Liver	GLSZM_SizeZoneNonUniformity, firstorder_Entropy
	Trachea	WL_LHL_GLRLM_GrayLevelNonUniformity, WL_HHL_GLSZM_LargeAreaHighGrayLevelEmphasis, WL_HLL_firstorder_Uniformity
	Adipose Tissue	GLCM_Idmn
Statistics Energy	Liver	GLRLM_HighGrayLevelRunEmphasis, WL_LLL_firstorder_Entropy
	Heart	firstorder_Entropy
	Air	WL_HLH_firstorder_10Percentile
Grey Level NonUniformity HLH	Liver	GLSZM_SizeZoneNonUniformity, GLCM_Imc2
	Heart	WL_HLH_GLSZM_ZoneEntropy, WL_LHH_GLSZM_LargeAreaEmphasis
LAA signature		
LAA950	Trachea	GLDM_GrayLevelVariance, LS_0_5_mm_3D_GLDM_DependenceNonUniformityNormalized, GLDM_DependenceNonUniformity, GLDM_LargeDependenceEmphasis
	Air	GLDM_LowGrayLevelEmphasis
	Spleen	LS_2_5_mm_3D_GLDM_DependenceVariance, GLDM_DependenceVariance
	Aorta	LS_3_5_mm_3D_GLDM_DependenceNonUniformityNormalized
LAA910	Trachea	LS_3_5_mm_3D_NGTDM_Complexity, LS_0_5_mm_3D_GLDM_LargeDependenceEmphasis
	Heart	LS_3_5_mm_3D_GLDM_GrayLevelNonUniformity, LS_0_5_mm_3D_GLDM_DependenceNonUniformity, LS_0_5_mm_3D_GLDM_LargeDependenceEmphasis

Table 3. Technome discovery. The qualified surrogates that led to the best predictive performance for one-year-survival (top) and centrilobular emphysema presence (bottom) are shown. CRs are ordered according to their explained variance of the respective feature. The sphericity of the Aerts was robust to technical variation and is thus omitted. Abbreviations for filters - LS: logSigma, WL: Wavelet. Further explanations about PyRadiomics classes' nomenclature such as 'GLSZM' can be found in the original publication[27].

Discussion

On CT images, all measurements should be seen as relative – a notion that is already accepted for bone mineral density measurements that are therefore calibrated by in-scan phantoms as seen, e.g., in the work by Kalender[36]. For bone mineral density, a calibration relates the measurements in a ROI to quantifications in the in-scan phantom. A univariate feature analysis without calibration thus appears dubious in most other cases as well. However, a machine learning algorithm can to some degree automatically calibrate for technical variation when predicting a label. We also observed this effect in our COPD study, where a naive approach that incorporated all available surrogates from non-lung CRs improved the predictive performance from 60.7% to 74.8%. To assist the classifier, we integrated an internal calibration based on surrogate qualification criteria in the training process of the classifier to enhance predictive performance. A feature signature indicative for the medical label is enhanced by additional incorporation of qualified surrogates. The qualification criteria for the surrogates stem from other sources of information usually used in radiomics feature stability analyses, such as phantom measurements, simulations but also best practices in statistics. For instance, our statistical qualification criterion enforces valid covariates to be used, as for a proper statistical assessment no interaction effect between the examined biological variable and the used covariate should occur. The introduced inSilico qualification criterion rewards the use of surrogates, when their association with the analysed feature *in vivo* can also be reidentified in simulation studies, which makes it highly unlikely that the association beween feature and surrogate is induced biologically. As in our case the technome classification uses a simple logistic regression model, the results can also be used for interpretable statistical assessment. The surrogates that are used within the model are appropriate for this task, as they have been shown to be suitable covariates by *in vivo*, *in vitro*, *in silico*, and orthogonality tests.

Our approach based on regularised models in which established imaging biomarker signatures are internally calibrated yielded better predictive performance than the signatures without calibration. Our models also outperformed conventional machine learning models with all available features and surrogates. This can be interpreted as an enhanced reproducibility of imaging biomarkers. Only an advanced deep learning approach[35] that employs a sparse autoencoder for dimensionality reduction performed better to predict one-year-survival. However, deep learning lacks the interpretability and probably also the reproducibility of our approach. This is especially important as the lack of interpretability and especially reproducibility[37] is already recognised as a large problem, not only for clinical studies, where a significant result of a potential biomarker associated with a treatment can often not be reproduced in similar studies. While the Imaging Biomarker Standardisation Initiative (IBSI)[38] addresses this problem by introducing unified feature definitions, the minimisation of technical variation remains a pressing issue.

First, the technome improved centrilobular emphysema prediction with a hard kernel and using the LAA signature from an AUC of 60.7% to 81.5%. Possibly, this could be of clinical value as a hard kernel was previously considered inadequate for emphysema assessment due to the inter-patient noise variation. The prediction of emphysema based on LAA features only with a soft kernel even showed a slightly worse performance of 77%. Second, by calibration, the Aerts signature to predict one-year-survival reproduced its expected predictive performance with an AUC of 66.5%. In the original study[20] this signature yielded a concordance index, which is a generalisation of an AUC, of 66–69% for unseen data. The prediction of one-year-survival based on the calibrated Aerts signature was superior in comparison to no calibration (49.2% via GLM and 57.6% via random forest), an optimised radiomics approach that uses all PyRadiomics features within the tumours and a nested cross-validation process (56.6%), and even comparable to the most advanced 'black-box' deep learning approaches (71.1%). Using a naive approach and entering the signature and all available surrogates in a random forest classifier led to an predictive performance that was even worse compared to no calibration (44.5% vs. 57.6%).

Despite not being our main focus, a second field of research was the explicit feature stabilisation in a phantom study. Although reducible technical variation can be read out via imaging protocol parameters, its information is sparse. For example the kernel is only a factor with no continuous parametrisation. This is also a limitation of the ComBat method, where for each imaging protocol a minimum number of cases must be available[12]. Therefore, it is not ensured that enough data for the respective imaging protocol can be collected. Also, a learned stabilisation, such as the deep learning approach by Jin[39], does of course not work for unseen scan and reconstruction parameters. In our phantom study, the calibration is applied on unseen scan and reconstruction parameters and therefore not even one sample for each imaging protocol can be used to learn the calibration. ComBat does by design not work in this scenario. The technome parametrises reducible and non-reducible technical variation in a unified manner via qualified surrogates, while the information which scan and reconstruction parameters are employed is not used. We compared the technome stabilisation to an adaption of the well-known RAVEL method[10], which is an advanced and well-performing MRI technique, that uses the singular vectors of variation in CRs to stabilise intensity distribution in MRI scans of the brain. The variance reduction performance of our method (90.4%) was numerically superior to RAVEL-like approach (76.3%) in the phantom study where object size, scan and reconstruction parameters were varied. This may be due to the fact that RAVEL uses the principal components of the CRs. Accordingly, each feature is calibrated by a selection from the same principal components of variation that are determined by a SVD of the CR's surrogates. The voxel value, in our case the feature value, is then linearly adjusted for those principal components of variation. From our phantom experiments however, it appears that each feature needs a more specific calibration that can profit from a surrogate qualification. The principal components may not always be appropriate, a feature that is affected by a cupping artifact may not be properly calibrated if the cupping artifact is not expressed in the principal components. Again, the combination of a qualification criterion and a regularised model performed better than a naive approach using all surrogates (90.4% vs. 76.7% variance reduction).

The technome discovery indicates that technical variation can have an impact on features, even when scan and reconstruction parameters are kept constant within a study. One finding of our experiments is that the best predictive performance was achieved when surrogates are used that are qualified by the inVivo and the inSilico qualification metric. The inVivo qualification rewards when surrogates are used that are correlated with the feature. The inSilico qualification however analyses if the found correlation between feature and surrogate can be reproduced in simulations of technical variation. For constant scan and reconstruction parameters, simulations even appear necessary to qualify appropriate surrogates. The finding that a high qualification of the surrogates translates to a higher predictive performance explains the good prediction results that were achieved by introducing the calibration loss in both clinical studies. As the best predictive performance was observed for highly qualified surrogates, it is very probable that the performance enhancement is due to the enforced internal calibration procedure and not due to integration of further biological information from the CRs. As a side effect we tested many different organs for imaging biomarkers, as the two studies were an exhaustive multi-organ analysis for COPD and mCRC. However, the addition of all available surrogates from all organs without qualification showed numerically inferior performance to the technome prediction. In the mCRC study, the integration even led to deterioration of predictive performance. We expect this to be a result of overfitting. Furthermore, the chosen CRs for calibration were plausible. The LAA signature measured in the parenchyma was internally calibrated by the representation of air within a cylinder in the trachea. For the mCRC data with contrast injection the technome used texture metrics within liver and heart. Interestingly, the technome chose the heart CR for calibration of contrast injection and not the aorta. This may be due to a more constant accumulation of contrast within the blood pool of the heart.

Our method has several limitations. The main limitation is the dichotomy of *predictive* and *stabilisation* mode. This case separation should ideally not be needed, as a perfect explicit calibration should also enhance predictive performance without knowledge of the medical label. Second, we did not calculate Kaplan-Meier curves for survival prediction, which would allow a better comparison of the Aerts signature's performance with the original study. An integration of the *predictive* mode within survival analysis would have needed further methodological effort. Third, we used only simple simulations and qualified them with the similarity argument of the inSilico qualification criterion. However, we expect an increase in predictive and stabilisation performance when more complex simulations are applicable, which was not yet possible due to the large computation time for each case. Fourth, our method in its current form is very computation-intensive and time-consuming. Especially the inSilico qualification needs to synthetically manipulate the images to collect associations between features and surrogates. Fifth, although we qualify surrogates for a feature by a variety of criteria, it is still possible that the non-uniform noise and artifact distribution in CT images can lead to the selection of suboptimal surrogates. The noise texture in the ROI can still significantly differ from the technically-induced texture that is found in the CR. Finally, of course, more data and clinical prediction tasks, e.g. for COPD[21], liver[40], or muscle diseases[41], are needed for further validation of the method. As a further proof of concept, the predictive performance of established imaging biomarkers should be reproduced in studies with larger technical variation.

In conclusion, we present a novel method that integrates qualification criteria for surrogates in the training of a classification or regression method leading to a predictive internal calibration. The method improved feature stabilisation in a phantom study, prediction of centrilobular emphysema in a COPD study and one-year-survival in mCRC study. The analysed studies indicate that the identification of surrogate qualification criteria and their integration in the training process of a prediction model is a promising field of research.

In the future, we plan to combine the surrogate qualification with ComBat and deep learning. For the deep learning variant, the calibration loss could be included as a regularisation term that controls the incorporation of CRs in the classification process. For ComBat, the substitution of the imaging protocol by qualified surrogates seems to be an interesting option.

Data availability

The relevant data supporting the findings are available within the article. The clinical study data are available from the corresponding author A.M. upon request.

Received: 14 March 2019; Accepted: 13 December 2019;
Published online: 24 January 2020

References

 1. Mulshine, J. L. *et al.* Role of the quantitative imaging biomarker alliance in optimizing ct for the evaluation of lung cancer screen–detected nodules. *J. Am. Coll. Radiol.* **12**, 390–395 (2015).
 2. Yasaka, K. *et al.* Precision of quantitative computed tomography texture analysis using image filtering: A phantom study for scanner variability. *Medicine* **96** (2017).
 3. Shafiq-ul Hassan, M. *et al.* Intrinsic dependencies of ct radiomic features on voxel size and number of gray levels. *Med. physics* **44**, 1050–1062 (2017).
 4. Larue, R. T. *et al.* 4dct imaging to assess radiomics feature stability: An investigation for thoracic cancers. *Radiother. Oncol.* **125**, 147–153 (2017).
 5. Lu, L., Ehmke, R. C., Schwartz, L. H. & Zhao, B. Assessing agreement between radiomic features computed for multiple ct imaging settings. *PloS one* **11**, e0166550 (2016).
 6. Kim, H. *et al.* Impact of reconstruction algorithms on ct radiomic features of pulmonary tumors: analysis of intra-and inter-reader variability and inter-reconstruction algorithm variability. *PloS one* **11**, e0164924 (2016).
 7. Ger, R. B. *et al.* Comprehensive investigation on controlling for ct imaging variabilities in radiomics studies. *Sci. Rep.* **8**, 13047 (2018).
 8. Kalra, M. K. *et al.* Techniques and applications of automatic tube current modulation for ct. *Radiol.* **233**, 649–657 (2004).
 9. Leek, J. T. *et al.* Tackling the widespread and critical impact of batch effects in high-throughput data. *Nat. Rev. Genet.* **11**, 733 (2010).
10. Fortin, J.-P. *et al.* Removing inter-subject technical variability in magnetic resonance imaging studies. *NeuroImage* **132**, 198–212 (2016).
11. Johnson, W. E., Li, C. & Rabinovic, A. Adjusting batch effects in microarray expression data using empirical bayes methods. *Biostat.* **8**, 118–127 (2007).
12. Orlhac, F., Frouin, F., Nioche, C., Ayache, N. & Buvat, I. Validation of a method to compensate multicenter effects affecting ct radiomics. *Radiol.* **291**, 53–59 (2019).
13. Kalender, W. A. Computed tomography: fundamentals, system technology, image quality, applications. *Comput. Tomogr. Fundamentals, Syst. Technol. Image Qual. Appl. by Willi A. Kalender, pp. 220. ISBN 3-89578-081-2. Wiley-VCH, Novemb. 2000.* **220** (2000).
14. Belhumeur, P. N., Hespanha, J. P. & Kriegman, D. J. Eigenfaces vs. fisherfaces: Recognition using class specific linear projection. Tech. Rep., Yale University New Haven United States (1997).
15. Dedeurwaerder, S. *et al.* A comprehensive overview of infinium humanmethylation450 data processing. *Briefings bioinformatics* **15**, 929–941 (2013).
16. Fortin, J.-P. *et al.* Functional normalization of 450 k methylation array data improves replication in large cancer studies. *Genome biology* **15**, 503 (2014).
17. Tabachnick, B. G. & Fidell, L. S. *Using multivariate statistics* (Allyn & Bacon/Pearson Education, 2007).
18. Pelikan, M., Goldberg, D. E. & Cantú-Paz, E. Boa: The bayesian optimization algorithm. In *Proceedings of the 1st Annual Conference on Genetic and Evolutionary Computation-Volume 1*, 525–532 (Morgan Kaufmann Publishers Inc., 1999).
19. Barnes, C. P., Silk, D., Sheng, X. & Stumpf, M. P. Bayesian design of synthetic biological systems. *Proc. Natl. Acad. Sci.* (2011).
20. Aerts, H. J. *et al.* Decoding tumour phenotype by noninvasive imaging using a quantitative radiomics approach. *Nat. Communic.* **5**, 4006 (2014).
21. Regan, E. A. *et al.* Genetic epidemiology of copd (copdgene) study design. *COPD: J. Chronic. Obstr. Pulm. Dis.* **7**, 32–43 (2011).
22. Martinez-Cantin, R. Bayesopt: A bayesian optimization library for nonlinear optimization, experimental design and bandits. *J. Mach. Learn. Res.* **15**, 3735–3739 (2014).
23. Pedregosa, F. *et al.* Scikit-learn: Machine learning in python. *J. Mach. Learn. Res.* **12**, 2825–2830 (2011).
24. Ding, C. & Peng, H. Minimum redundancy feature selection from microarray gene expression data. *J. Bioinform. Comput. Biol.* **3**, 185–205 (2005).
25. Ramírez-Gallego, S. *et al.* Fast-mrmr: Fast minimum redundancy maximum relevance algorithm for high-dimensional big data. *Int. J. Intell. Syst.* **32**, 134–152 (2017).
26. Keil, S. *et al.* Semi-automated quantification of hepatic lesions in a phantom. *Investig. Radiol.* **44**, 82–88 (2009).
27. van Griethuysen, J. J. *et al.* Computational radiomics system to decode the radiographic phenotype. *Cancer Res.* **77**, e104–e107 (2017).
28. Moltz, J. H. *et al.* Advanced segmentation techniques for lung nodules, liver metastases, and enlarged lymph nodes in ct scans. *IEEE J. Select. Top. Signal Process.* **3**, 122–134 (2009).
29. Rémy-Jardin, M. *et al.* Detection and phenotyping of emphysema using a new machine learning method. In *RSNA 2018* (RSNA, 2018).
30. Seifert, S. *et al.* Hierarchical parsing and semantic navigation of full body ct data. In *Medical Imaging 2009: Image Processing*, vol. 7259, 725902 (International Society for Optics and Photonics, 2009).
31. Yang, D. *et al.* Automatic liver segmentation using an adversarial image-to-image network. In *International Conference on Medical Image Computing and Computer-Assisted Intervention*, 507–515 (Springer, 2017).
32. Kuhnigk, J.-M. *et al.* Morphological segmentation and partial volume analysis for volumetry of solid pulmonary lesions in thoracic ct scans. *IEEE Transact. Med. Imag.* **25**, 417–434 (2006).
33. Nörenberg, D. *et al.* Deep learning based radiomics and its usage in prediction for metastatic colorectal cancer. In *RSNA 2018* (RSNA, 2018).
34. Wels, M., Lades, F., Mühlberg, A. & Sühling, M. General purpose radiomics for multi-modal clinical research. In *Proc. SPIE Medical Imaging: Computer-Aided Diagnosis, San Diego, CA, USA, Feb 2019* (SPIE, 2019).

35. Katzmann, A. *et al.* Predicting lesion growth and patient survival in colorectal cancer patients using deep neural networks. In *International conference on Medical Imaging with Deep Learning* (Springer, 2018).
36. Kalender, W. A. *et al.* The european spine phantom—a tool for standardization and quality control in spinal bone mineral measurements by dxa and qct. *Eur. J. Radiol.* **20**, 83–92 (1995).
37. Baker, M. 1,500 scientists lift the lid on reproducibility. *Nat. News* **533**, 452 (2016).
38. Zwanenburg, A., Leger, S., Vallières, M. & Löck, S. Image biomarker standardisation initiative. *arXiv preprint arXiv:1612.07003* (2016).
39. Jin, H. & Kim, J. H. Deep learning-enabled scan parameter normalization of imaging biomarkers in low-dose lung ct. In *2018 International Workshop on Advanced Image Technology (IWAIT)*, 1–2 (IEEE, 2018).
40. Hayashi, M. *et al.* Correlation between the blood supply and grade of malignancy of hepatocellular nodules associated with liver cirrhosis: evaluation by ct during intraarterial injection of contrast medium. *AJR. Am. J. Roentgenol.* **172**, 969–976 (1999).
41. Mühlberg, A. *et al.* Three-dimensional distribution of muscle and adipose tissue of the thigh at ct: Association with acute hip fracture. *Radiology* 181112 (2018).

Acknowledgements

We thank Grzegorz Soza, Felix Durlak, Andreas Wimmer (Siemens Healthineers, Forchheim, Germany), Andreas Friedberger, Oleg Museyko, Klaus Engelke, Oliver Chaudry (Institute of Medical Physics, FAU Erlangen-Nürnberg, Germany), Jan Moltz (Fraunhofer MEVIS, Bremen, Germany), and Wolfgang Goldmann (Biophysics Group, FAU Erlangen-Nürnberg, Germany) for valuable discussions. This project has received funding from the German Federal Ministry of Education and Research as part of the PANTHER project und grant agreement no. 13GW0163A.

Author contributions

A.M. developed the theory, implemented the algorithms and conducted all analyses, F.L. and R.K. extracted the radiome for mCRC and COPD data, A.K., O.T., M.W., R.K. and M.S. consulted with theory development, F.L. implemented the mRMR algorithm in the software, M.S. and M.W. consulted in choosing suitable algorithms for control region segmentation, D.N., T.H. and S.M. screened, collected and annotated the mCRC image data. J.H., V.H. designed the mCRC study, J.H. annotated the clinical mCRC data. J.F. and M.R.J. designed the COPD study, screened, collected and annotated the clinical and image COPD data. All authors reviewed the manuscript.

Competing interests

A.M., A.K., R.K., M.W., F.L. and M.S. work for Siemens Healthineers. D.N., T.H., S.M., J.F. and M.R.J. declare no competing interests. J.H. served on advisory board for Roche, has received honoraria from Roche and travel support from Novartis, V.H. has received honoraria from Merck KgaA, Roche A.G., Amgen, Sanofi, Sirtex and Baxalta and has received travel support from Merck KgaA, Roche A.G., Amgen, Sirtex and Baxalta and has served on advisory boards for Merck KgaA, Roche A.G., Amgen, Sanofi, Lilly, Sirtex, Böhringer Ingelheim, Baxalta, Taiho and Merrimack.

Additional information

Supplementary information is available for this paper at https://doi.org/10.1038/s41598-019-57325-7.

Correspondence and requests for materials should be addressed to A.M.

Reprints and permissions information is available at www.nature.com/reprints.

Publisher's note Springer Nature remains neutral with regard to jurisdictional claims in published maps and institutional affiliations.

European Radiology
https://doi.org/10.1007/s00330-020-07192-y

GASTROINTESTINAL

The relevance of CT-based geometric and radiomics analysis of whole liver tumor burden to predict survival of patients with metastatic colorectal cancer

Alexander Mühlberg[1] · Julian W. Holch[2] · Volker Heinemann[2] · Thomas Huber[3,4] · Jan Moltz[5] · Stefan Maurus[4] · Nils Jäger[4] · Lian Liu[2] · Matthias F. Froelich[3,4] · Alexander Katzmann[1] · Eva Gresser[4] · Oliver Taubmann[1] · Michael Sühling[1] · Dominik Nörenberg[3,4] 🄳

Received: 11 March 2020 / Revised: 2 July 2020 / Accepted: 13 August 2020
© European Society of Radiology 2020

Abstract

Objectives To investigate the prediction of 1-year survival (1-YS) in patients with metastatic colorectal cancer with use of a systematic comparative analysis of quantitative imaging biomarkers (QIBs) based on the geometric and radiomics analysis of whole liver tumor burden (WLTB) in comparison to predictions based on the tumor burden score (TBS), WLTB volume alone, and a clinical model.

Methods A total of 103 patients (mean age: 61.0 ± 11.2 years) with colorectal liver metastases were analyzed in this retrospective study. Automatic segmentations of WLTB from baseline contrast-enhanced CT images were used. Established biomarkers as well as a standard radiomics model building were used to derive 3 prognostic models. The benefits of a geometric metastatic spread (GMS) model, the Aerts radiomics prior model of the WLTB, and the performance of TBS and WLTB volume alone were assessed. All models were analyzed in both statistical and predictive machine learning settings in terms of AUC.

Results TBS showed the best discriminative performance in a statistical setting to discriminate 1-YS (AUC = 0.70, CI: [0.56, 0.90]). For the machine learning–based prediction for unseen patients, both a model of the GMS of WLTB (0.73, CI: [0.60, 0.84]) and the Aerts radiomics prior model (0.76, CI: [0.65, 0.86]) applied on the WLTB showed a numerically higher predictive performance than TBS (0.68, CI: [0.54, 0.79]), radiomics (0.65, CI: [0.55, 0.78]), WLTB volume alone (0.53, CI: [0.40. 0.66]), or the clinical model (0.56, CI: [0.43, 0.67]).

Conclusions The imaging-based GMS model may be a first step towards a more fine-grained machine learning extension of the TBS concept for risk stratification in mCRC patients without the vulnerability to technical variance of radiomics.

Key Points
• *CT-based geometric distribution and radiomics analysis of whole liver tumor burden in metastatic colorectal cancer patients yield prognostic information.*
• *Differences in survival are possibly attributable to the spatial distribution of metastatic lesions and the geometric metastatic spread analysis of all liver metastases may serve as robust imaging biomarker invariant to technical variation.*
• *Imaging-based prediction models outperform clinical models for 1-year survival prediction in metastatic colorectal cancer patients with liver metastases.*

Alexander Mühlberg and Julian W. Holch contributed equally to this work.

Electronic supplementary material The online version of this article (https://doi.org/10.1007/s00330-020-07192-y) contains supplementary material, which is available to authorized users.

✉ Dominik Nörenberg
    Dominik.Noerenberg@medma.uni-heidelberg.de

[1] CT R&D Image Analytics, Siemens Healthineers, Forchheim, Germany

[2] Comprehensive Cancer Center Munich, University Hospital, LMU Munich, Munich, Germany

[3] Department of Radiology and Nuclear Medicine, University Medical Center Mannheim, Medical Faculty Mannheim, Heidelberg University, Mannheim, Germany

[4] Department of Radiology, Munich University Hospitals, Munich, Germany

[5] Fraunhofer Institute for Medical Image Computing MEVIS, Bremen, Germany

Published online: 27 August 2020

🌻 Springer

Keywords Colorectal cancer · Tumor burden · Radiomics · Machine learning · Spatial analysis

Abbreviations

1-YS	1-Year survival
ARP	Aerts radiomics prior
CPH	Cox proportional hazards
CRLM	Colorectal liver metastases
GMS	Geometric metastatic spread
LLD	Liver-limited disease
mCRC	Metastatic colorectal cancer
MS	Metastatic spread
MSx(y/z)	Metastatic spread along CT scanner x(y/z)-axis
PTS	Primary tumor sidedness
QIB	Quantitative imaging biomarker
TBS	Tumor burden score
WLTB	Whole liver tumor burden

Introduction

Colorectal cancer is the third most common cancer worldwide [1]. Approximately 50% of patients with colorectal cancer will be diagnosed with metastases either at the time of diagnosis or as part of recurrent disease, whereas the liver is the most common site for metastases [1]. Although surgical resection of hepatic metastases is considered the only curative treatment option, approximately 85% of these patients are ineligible for this treatment due to large tumor burden, multifocal disease, or inadequate liver function [2, 3]. Computed tomography (CT) provides valuable capabilities for non-invasive assessment and quantification of colorectal liver metastases (CRLM) towards the development of predictive quantitative imaging biomarkers (QIBs) [4–6]. In recent years, there has been an increased interest to understand survival and response to therapy in tumor patients using the whole tumor burden rather than single lesions [7, 8]. For CRLM patients, the volume of the whole liver tumor burden (WLTB) and the tumor burden score (TBS) were quantified. The TBS is the Pythagorean addition of the lesion number and the diameter of the largest lesion. This measurement was capable to better estimate survival than the number of lesions or the diameter of the largest lesion alone [9, 10]. Although being a natural extension of this concept, the relevance of geometric measures of the WLTB distribution such as distances between various lesions has not yet been evaluated.

Furthermore, texture analysis and machine learning [4–6, 11, 12] are playing an increasingly important role in radiology, displacing statistical analysis of QIB. A special branch of this research represents radiomics. This is based on extracting a large number of quantitative features from the images and combining them with machine learning to make the diagnosis, therapy response, and outcome prediction more accurate [13,

14]. In patients with CRLM, radiomics analysis of target lesions was shown to significantly correlate with response to chemotherapy, as well as with survival [4–6, 12].

Since the added predictive value of the geometric or radiomics analysis of WLTB is not known, we compare the predictive performance of established clinical and quantitative imaging biomarkers and novel exploratory whole liver tumor burden–based QIBs in CRLM patients by a statistical and also a machine learning approach.

The purpose of our study is therefore to investigate the prediction of 1-year survival in patients with metastatic colorectal cancer with use of a systematic comparative analysis of QIBs based on the geometric and radiomics analysis of WLTB in comparison to predictions based on the TBS, WLTB volume alone, and a clinical model.

Materials and methods

Our retrospective study was approved by and registered with the local institutional review board of the Ludwig-Maximilians-University Munich (approval number: 502-16). Written informed consent was obtained from all subjects.

Study sample

A database of patients with metastatic colorectal cancer from January 2007 to October 2017 was reviewed and 485 patients with metastatic colorectal cancer were identified. Of the 485 enrolled oncological patients, 269 had an available baseline CT scan. Of those, for 220 patients, sufficient clinical data were available. A total of 133 patients of this cohort had colorectal liver metastases (CRLM). Further 15 patients had to be excluded due to limited image quality, native scans, or motion artifacts. Additionally, 15 further patients had to be excluded due to missing information regarding survival status or lack of clinical follow-up information. The final study cohort therefore consisted of 103 patients, of which 82 survived at least 1 year. A flowchart describing the exclusion criteria for patients can be found in Fig. 1.

Imaging studies

Our retrospective study includes baseline CT scans which were acquired using a variety of multidetector-row CT scanners from different manufacturers (see Table 1); default settings were 120 kV tube voltage. Weight-adapted contrast agent was applied intravenously, and images were acquired in portal venous phase and reconstructed using a standard soft tissue kernel. Slice thickness varied between 0.75 and 5 mm.

Fig. 1 Flowchart of patient exclusion resulting in the final study cohort of 103 CRLM patients

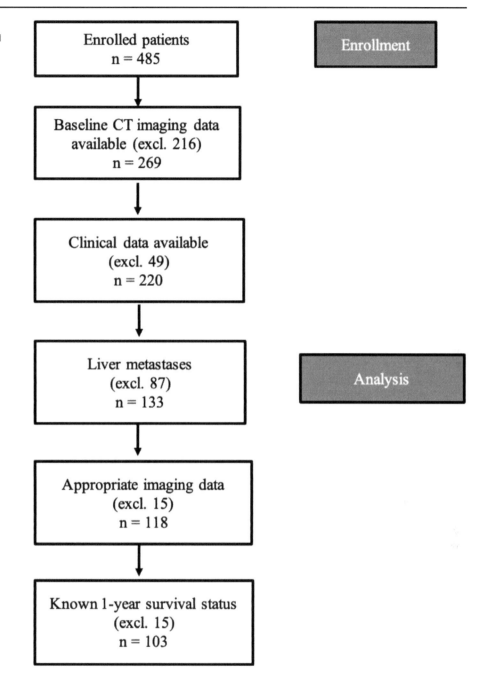

WLTB segmentations

All CT scans were reviewed independently in a randomized fashion and blinded to the clinical data; only contrast-enhanced CT scans were used for further analysis. All included CT scans were screened for metastases by 2 board-certified radiologists (D.N. and T.H.) with each having 6 years of experience in oncological imaging. WLTB segmentations were performed automatically using custom software based on MeVisLab (MeVis Medical Solutions, Fraunhofer MEVIS) with support of a convolutional neural network [15]. If needed, both radiologists could revise the WLTB segmentations interactively by adding tumors or redefining tumor contours.

 Springer

Table 1 Baseline characteristics of 103 included CRLM patients and used CT scanner types

Characteristic	Value
No. of patients	103
Age (years)	61.0 ± 11.2
Female sex	48 (46.6)
Liver-limited disease (LLD)	44 (42.7)
Syn-/metachronous disease	synchronous: 81 (78.6); metachronous: 22 (21.4)
Whole liver tumor burden (WLTB) volume (cm^3)	332.7 ± 469.7
Primary tumor sidedness (PTS)	right: 34 (41.2) left: 68 (58.8) ns: 1
1-year survival (1-YS)	yes: 82 (79.6) no: 21 (20.4)
CT scanner type	1: 1, 2: 4, 3: 1, 4: 9, 5: 1, 6: 2, 7: 1, 8: 1, 9: 5, 10: 1, 11: 1, 12: 2, 13: 4, 14: 1, 15: 1, 16: 2, 17: 1, 18: 1, 19: 1, 20: 1, 21: 2, 22: 39, 23: 4, 24: 2, 25: 11, 26: 1, ns: 2

Data in parentheses are percentages. *ns* not specified. Continuous variables are given as mean ± SD. 1: Alexion, 2: Aquilion, 3: Asteion, 4: Biograph 64, 5: BrightSpeed S, 6: Brilliance 16P, 7: Brilliance 64, 8: Definition AS, 9: Definition AS+, 10: Discovery 690, 11: Discovery CT750 HD, 12: Emotion 16, 13: Emotion 16 (2007), 14: Emotion 16 (2010), 15: Emotion 6, 16: Emotion Duo, 17: Ingenuity CT, 18: MX 16-slice, 19: SOMATOM Definition AS, 20: SOMATOM Definition AS+, 21: SOMATOM Definition Edge, 22: SOMATOM Definition Flash, 23: SOMATOM Force, 24: Sensation 16, 25: Sensation 64, 26: iCT 256

Analyzed prognostic models

A general overview of a radiomics workflow can be found in Fig. 2a. After image segmentations, imaging features were extracted by an in-house software that also integrates the PyRadiomics library [16]. Features were grouped into 5 (i to v) prognostic models (Fig. 2b): (i) the imaging prior model, (ii) the clinical prior model, (iii) the Aerts radiomics prior (ARP) model, (iv) the geometric metastatic spread (GMS) model, and (v) the naive model. An overview of the models is shown in Fig. 2b. In detail, 3 of them (i–iii) are based on prior knowledge, one (iv) is our own hypothesis to introduce a novel quantitative imaging biomarker related to the spatial tumor distribution of all liver metastases, and the last model (v) uses all available features in a mechanic standard radiomics model building approach. The prior models are grouped according to their source, e.g., imaging or clinical data.

At first, as a benchmark, we propose an imaging prior (i) model. This model incorporates already discovered discriminative quantitative imaging biomarkers (QIBs) found in oncological imaging such as TBS [9, 17], primary tumor sidedness (PTS) [18–20], tumor attenuation [21], and also the whole liver tumor burden volume [22, 23]. PTS was defined as right-sided or left-sided if the tumor arose from the cecum to the hepatic flexure or from the splenic flexure to the rectum, respectively. Those QIBs are also analyzed individually for their prognostic value. In analogy to the imaging prior model, a (ii) clinical prior model is proposed based on current clinical parameters, including data from laboratory and

histopathology analysis. This model consists of PTS [19, 20, 22], presence of liver-limited disease (LLD) [20], age, sex, grading, syn-/metachronous metastases [20, 24], histology, and carcinoembryonic antigen (CEA) levels as well as UICC and TNM staging [20]. LLD denotes a specific subgroup of metastatic colorectal cancer patients where the liver is the only metastatic site.

Furthermore—since this model provided good predictive performance when applied on target lesions in multiple oncological imaging studies—we also evaluate the established Aerts radiomics prior (ARP) model (iii) [13, 25–27] for the WLTB analysis. This model consists of 4 quantitative image features describing the tumor heterogeneity and compactness.

As described above, we hypothesize that the spatial geometric distribution of the tumors within the liver may also be of diagnostic value and propose a (iv) geometric metastatic spread (GMS) model. This model consists of the maximum distance of liver metastases along x, y, and z CT scanner axes, termed metastatic spread (MSx, MSy, MSz) and the addition of these squared distances (MS). Furthermore, the GMS model integrates a dispersion quantification by 2 features, namely the surface-area-to-volume ratio (SA/V) and the compactness of the spatial metastases distribution. A formal description of the model can be found in Suppl. Mat. B.

Finally, a mechanic construction of a predictive model based on all extractable features (PyRadiomics library + imaging priors + clinical priors + ARP + GMS) solely by machine learning and a minimum redundancy maximum relevance (mRMR) feature selection [28] is tested and termed (v) naive approach. The term naive indicates that no prior knowledge or intuition was used which was formed on the

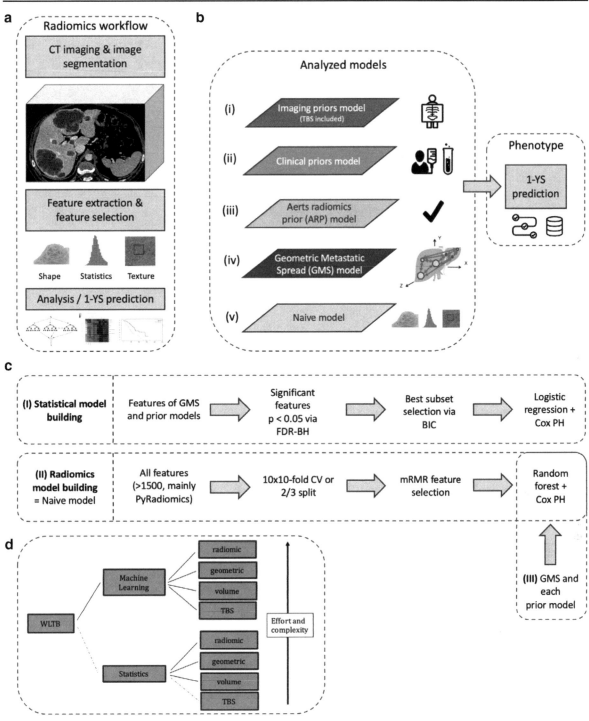

Fig. 2 Basic schemes of our analyses pipelines. **a** The general radiomics workflow. **b** The analyzed prior models (i–iii), the GMS model (iv), and the standard radiomics model building (naive, v). **c** The statistical (I) as well as standard radiomics machine learning (II) model building, and also our machine learning setup based on the GMS hypothesis and prior models (III). **d** The complexity and effort involved in the respective analyses. Roughly divided, the complexity and/or effort associated with each analysis increases from bottom to top, due to a higher effort to generate WLTB segmentations, the higher model complexity of a non-linear machine learning approach in comparison to a regularized statistical model and the complexity of controlling the impact of scan parameter variation on texture measurements within a radiomics analysis

basis of previous study results but a standard radiomics model building process was pursued. Figure 2d describes and ranks the complexity and effort involved in the respective analyses described above.

Data analysis

1-YS and survival time were measured from the date of initial baseline CT at time of the initial diagnosis of metastatic disease to the date of death (if applicable). An overview of the data analysis pipeline can be found in Fig. 2c displaying both our statistical (I) as well as machine learning (II, III) model building. As some biomarkers are rather analyzed statistically while others such as the ARP model are derived from machine learning approaches, we had to include both analyses for a profound comparison.

For statistical model building (Fig. 2c (I)), descriptive statistics, such as means and standard deviation (SD) for continuous variables and frequencies for categorical variables, are used to summarize the data and each feature of prior models and GMS model introduced above. Due to multiple testing corrections, the naive model using all available features (> 1500) is not suitable for statistical analysis. Our statistical pipeline approach is similar to prior studies [29, 30]. Univariable statistics are reported by p value determined via Student's t test if applicable (Shapiro-Wilk and Levene's test) or Wilcoxon's rank-sum test for continuous variables, a Fisher's exact test for categorical variables with 2 factors, and a chi-square test for > 2 factors. A two-sided p value < 0.05 was considered significant. For multivariable model construction, univariably significant features are selected after false discovery rate Benjamini-Hochberg multiple testing correction. This feature selection is further reduced by a best subset selection according to the Bayesian information criterion. Multivariable models are then fitted by logistic regression for the five best subsets and reported by the statistical AUC with 95% confidence interval (CI) and odds ratios (OR) for features normalized to 1 SD with 95% CI. Additionally, univariable and multivariable Cox proportional hazard (CPH) models are used to determine concordance index (C-index, a generalization of the AUC applicable for survival regression). For the multivariable model, survival differences based on the fitted CPH median survival stratification in high- and low-risk groups are visualized by Kaplan-Meier curve and survival differences are quantified by the log-rank test. For a deeper understanding of the features, a univariable Spearman correlation heatmap with absolute values and dendrogram is generated to quantify associations of clinical with imaging variables.

For the machine learning approach (Fig. 2c (II), (III)), two methods are used to generate derivation and validation data: a temporal 2/3 split, i.e., patients are split according to the date of their baseline scan, and 10 × 10-fold cross-validation (CV),

i.e., 10 different random seeds are used for a 10-fold CV. 10 × 10 CV is used for the prediction of 1-YS. The temporal split is introduced for the survival regression to estimate temporal batch effects on the prediction, e.g., a temporal change of doctor in charge, a common effect described by Leek et al [31]. A random forest for 1-YS prediction and a CPH for survival prediction are trained on the derivation data for each of the introduced prognostic models. Predictive performance is evaluated by predictive AUC (random forest), C-index (survival regression), and significance for the models on the validation data. Additionally, a CPH median survival stratification threshold is determined on the derivation data and applied to the validation data. Then, Kaplan-Meier curves are generated and the log-rank test is used to assess the significance of the predicted risk group on the validation data. Data analysis is done with R (version 3.3.2, www.R-project.org) and Lifelines [32].

Results

Demographic data

Demographics of the included patients and used CT scanner types of our study sample are shown in Table 1.

WLTB segmentations

In Fig. 3, four representative patients are shown visualizing patients with varying values for TBS, WLTB volume, and geometric metastatic spread.

Data analysis and survival prediction

Significant features of the univariable statistical evaluation before multiple testing correction are shown in Fig. 4a. The complete results of univariable analysis are shown in Table 1S of Suppl. Mat. A.

Using univariable statistics, features with the significant discriminative performance were TBS, MS, MSy, MSz, and the compactness of the tumor distribution. The GMS features MS, MSy, MSz, and compactness also showed a significant goodness of fit and C-index between 0.62 and 0.65.

The association heatmap with dendrogram to visualize the univariable associations of imaging and clinical variables is shown in Fig. 5. The highest association between clinical parameters and imaging was found between CEA and WLTB volume and the metastatic spread along the CT y-axis MSy and the M staging.

Regarding multivariable statistics, of the 5 significant features, only the TBS remains in the best multivariable model according to the Bayesian information criterion (Fig. 4b, c;

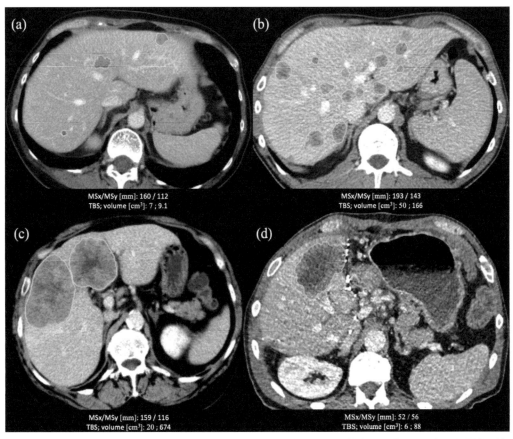

MSx/MSy [mm]: 160 / 112
TBS; volume [cm³]: 7 ; 9.1

MSx/MSy [mm]: 193 / 143
TBS; volume [cm³]: 50 ; 166

MSx/MSy [mm]: 159 / 116
TBS; volume [cm³]: 20 ; 674

MSx/MSy [mm]: 52 / 56
TBS; volume [cm³]: 6 ; 88

Fig. 3 Exemplary WLTB segmentations of four patients with CRLM and their according measurements (MSx/MSy [mm]; TBS; volume [cm³]). **a** A patient with intermediate metastatic spread, low TBS, and low WLTB volume and "no 1-year survival" (1-YS). **b** A patient with high metastatic spread and TBS, intermediate volume, and also "no 1-YS." Patients in **c** and **d** had "1-YS" with intermediate (**c**) or low (**d**) TBS and intermediate (**c**) or low (**d**) metastatic spread while their tumor volume was larger than that in **a**. Patient **a** appears to be especially interesting, as "no 1-YS" is correctly indicated here by the metastatic spread while the TBS points rather towards "1-YS"

Table 2) and shows a good discriminative performance with a discriminative AUC of 0.70 [0.56, 0.90] for 1-YS. However, the best 3 models achieve similar performance. All 5 multivariable models consist of only one feature.

Results for the machine learning analysis to predict 1-YS for unseen data are shown in Table 3 and Fig. 6a. Here, the QIB TBS achieved also a good predictive performance for 1-YS prediction with a predictive AUC of 0.68 [0.54, 0.79]. The QIBs WLTB volume, attenuation of WLTB, and PTS individually showed inferior performance with AUCs between 0.5 and 0.57. The imaging prior model (i) consisting of all QIBs yields similar results as TBS alone with a predictive AUC of 0.67 [0.54, 0.79], whereas the clinical prior model (ii) achieves 0.56 [0.43, 0.67]. A combination of both prior models achieves again a similar performance with 0.66 [0.54, 0.77] (data not shown). The GMS model (iv) and the ARP model (iii) were numerically superior to both with a predictive AUC of 0.73 [0.602, 0.84] and with 0.76 [0.65, 0.86], respectively. The naive model, i.e., the standard radiomics model

building approach using all features, results in an AUC of 0.65 [0.55, 0.78], highlighting the importance of prior knowledge or intuition.

Kaplan-Meier curves for the predictive performance on unseen data are shown in Fig. 6b. C-index was highest for the GMS and the ARP model with 0.70 and 0.66. Again, the TBS showed a good performance with a C-index of 0.64.

Discussion

We investigated whether whole liver tumor burden (WLTB), and especially geometric and radiomics analyses of WLTB, extracted from pretreatment CT, could be used as prognostic biomarkers of the 1-YS of patients with colorectal liver metastases (CRLM). We compared established QIB and five different models ((i) imaging prior, (ii) clinical prior, (iii) Aerts radiomics prior (ARP), (iv) geometric metastatic spread (GMS), (v) naive model, i.e., standard radiomics model

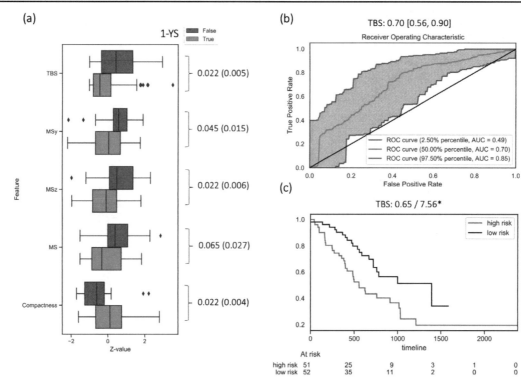

Fig. 4 Statistical analyses to assess 1-YS and survival time. **a** Univariable significant features are shown by boxplots and *p* values with (without) multiple testing correction. **b** AUC with CI for the best subset multivariable model consisting of only TBS. **c** Kaplan-Meier curve of the multivariable model for high- and low-risk groups. Results in **c** are given as model C-index/score of log-rank test (*$p < 0.05$, **$p < 0.01$). *TBS* tumor burden score, *MS* metastatic spread, *MSx* metastatic spread along CT scanner x-axis)

building), for predicting 1-YS. We therefore analyzed contrast-enhanced CT scans of 103 patients with CRLM scheduled for first-line therapy to assess the prognostic value of each model. Our goal was a systematic comparative analysis of quantitative imaging biomarkers applied on WLTB in metastatic colorectal cancer patients and if potential and robust predictors of patient survival could be identified, which may serve as early imaging biomarkers for risk stratification. The main findings of our study are that geometric but also radiomics WLTB-based measures are significantly associated with the outcome of patients with mCRC. The tumor burden score represents a reliable predictive QIB and shows higher predictive values than all clinical models or the WLTB volume alone. However, the ARP as well as GMS model even shows a numerically higher predictive performance than TBS in a machine learning setting.

Generally, heterogeneity information using radiomics and texture analysis are achieved for one target lesion of a single anatomical site. In prior cancer studies, the prognostic utility of radiomics was used for survival prediction and disease relapse in head and neck as well as lung cancer patients [13, 33–35]. A comparable radiomics approach was also used to predict survival and therapy response in patients with nasopharyngeal cancer [36] or glioblastoma [37], the latter based

on MRI data. Aerts et al [13] defined a four-feature signature, which represents the ARP model of our study, by focusing on the most robust features for prognostication in a lung dataset, and validated their signature using independent lung and head and neck cancer patient cohorts. Additionally, several previous studies have also analyzed gross whole tumor morphology, including tumor size and number, as important QIBs for survival prediction in mCRC patients [7–10, 38]. In terms of CRLM, the tumor burden score, incorporating maximum tumor size and number of lesions, was analyzed for survival discrimination in mCRC patients [9] and was outlined as an accurate tool to account for the impact of tumor morphology on long-term survival. As shown previously, TBS-based survival analysis revealed excellent prognostic discrimination for the TBS model and outperformed discrimination based on maximum tumor size and/or total number of lesions as performed in daily clinical routine by use of the established Fong score [9]. Our study therefore assesses and compares the value of a geometric metastatic spread model and radiomics analysis with the TBS as an already established predictive QIB. In our study, TBS reproduced its strong discriminative performance in a regularized logistic regression statistics approach, i.e., model fit and application on the same data, but the GMS and the ARP model in combination with a random forest

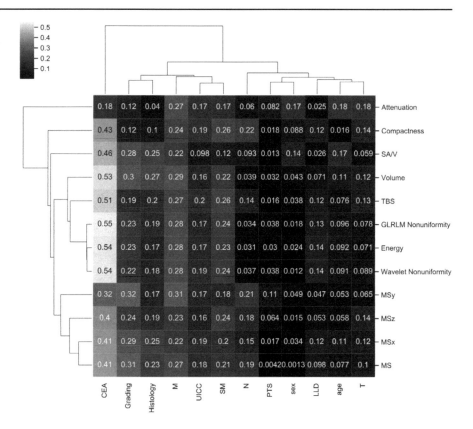

Fig. 5 Spearman correlation heatmap with absolute values and dendrogram to visualize association between imaging and clinical variables. *SM* syn-/metachronous disease, *PTS* primary tumor sidedness, *TBS* tumor burden score, *MS* metastatic spread, *MSx* metastatic spread along CT scanner x-axis

classifier yielded an enhanced predictive performance for unseen data. The performance of the ARP model appears plausible since this model has often been shown to be of high and reproducible predictive value across various cancer types [13, 16, 25, 26]. Of note, a prior study targeting the vulnerability of radiomics approaches determined that the tumor volume alone in the aforementioned head and neck and lung cancer datasets [13, 33–35] has a similar prognostic accuracy as the ARP model [39]. The authors conclude that the ARP model was a surrogate for tumor volume and that intensity and texture values were not pertinent for prognostication [40]. In our study, we analyzed both the ARP model and the volume only approach for prognostication and could clearly show a higher predictive value of the ARP model in a CRLM dataset. To

overcome underlying dependencies of intensity and texture-based measures, our study outlines the GMS of liver metastases as newly defined imaging biomarker with a comparable prognostic accuracy, but invariant to technical variation and independent of texture- or intensity-based values. The spread of metastases as quantified by the GMS could be expected to be diagnostically relevant as it might influence the resectability of liver metastases based on their spatial distribution. The GMS model shows strong performance in classification and survival regression and its features were also significant in a univariable statistics analysis.

Thus, our additional flexible machine learning approach using the geometry of the spatial WLTB distribution as well as the Aerts radiomics prior model led to a numerically

Table 2 Multivariable statistical model (logistic regression) with odds ratio (OR) and CI, *p* value, and the AUC with CI

Model	BIC	Multivariable model	OR [CI]	*p*	AUC discrimination
N1	105.0	TBS	1.9 [1.2, 3.1]	0.004	0.70* [0.57, 0.83]
N2	105.5	MSz	2.1 [1.2, 3.6]	0.008	0.69* [0.55, 0.82]
N3	106.0	Compactness	2.1 [1.2, 3.9]	0.012	0.71* [0.58, 0.83]
N4	107.4	MS	1.8 [1.1, 3.0]	0.02	0.66* [0.53, 0.79]
N5	108.5	MSy	1.8 [1.0, 3.2]	0.04	0.67* [0.53, 0.81]

TBS tumor burden score. *Significantly (*p* < 0.05) better than informed guessing. All 103 patients were used for model construction; i.e., no data is unseen (discrimination)

Table 3 Established QIB and models in the machine learning analysis to predict 1-year survival for unseen data

Feature	Variable importance [%] CV	AUC CV	C-index SPLIT Train (N = 69)	Test (N = 34)
mCRC QIBs				
TBS[#]	–	0.68* [0.54, 0.79]	0.66	0.64
PTS[#]	–	0.51 [0.37, 0.66]	0.58	0.59
WLTB volume[#]	–	0.53 [0.40, 0.66]	0.62	0.50
Attenuation[#]	–	0.57 [0.44, 0.69]	0.51	0.41
Imaging priors model				
TBS	31.4	0.67*[0.54, 0.79]	0.69	0.63
PTS	7.1			
WLTB volume	31.4			
Attenuation	30.1			
Clinical priors model				
PTS	5.2	0.56 [0.43, 0.67]	0.65	0.49
Sex	3.5			
Age	25.1			
CEA	20.1			
Grading	8.1			
Histology	1.8			
Syn-/metachronous disease	2.4			
UICC	10.6			
LLD	2.6			
T	6.2			
M	5.3			
N	9.1			
Aerts radiomics prior (ARP) model				
Energy	21.1	0.76* [0.65, 0.86]	0.66	0.67
Compactness	32.8			
GLRLM non-uniformity	25.7			
Wavelet non-uniformity	20.3			
Geometric metastatic spread (GMS) model				
MSx (mm)	16.4	0.73* [0.60, 0.84]	0.66	0.70
MSy (mm)	14.4			
MSz (mm)	18.0			
MS (mm)	15.5			
SA/V	14.3			
Compactness	21.4			
Naïve model				
§		0.65* [0.55, 0.78]	0.68	0.48

The prediction models are based on a random forest. Results are given, if applicable, with their feature importance for the 10-fold CV (CV) and the temporal 2/3 split (SPLIT). AUC is shown with 95% confidence interval. [#] Logistic regression is used for prediction; *Significantly ($p < 0.05$) better than informed guessing; [§] Variable importance not meaningful, due to shared importance of correlated features and distinct important features of each fold. *QIB* quantitative imaging biomarker, *PTS* primary tumor sidedness, *TBS* tumor burden score, *GMS* Geometric metastatic spread, *ARP* Aerts radiomics prior

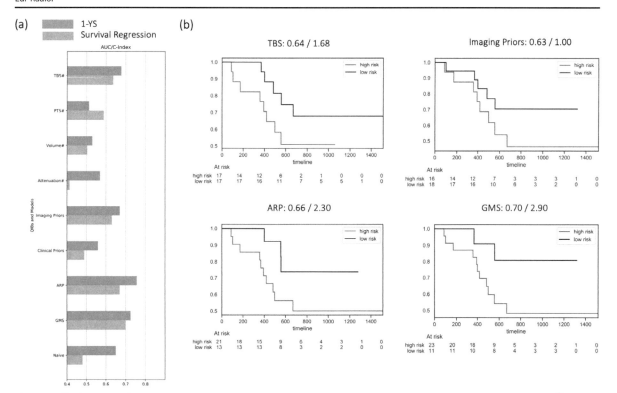

Fig. 6 Machine learning analyses to predict 1-YS and survival time for unseen patients. **a** AUC for 10 × 10 CV (light gray) and C-index (red) for temporal 2/3 split. **b** Selected Kaplan-Meier curves for unseen patients and predictive performance in temporal 2/3 split. Results in **b** are given as model: C-index/score of log-rank test (*p < 0.05, **p < 0.01). [#]1-YS prediction based on a logistic regression model. *PTS* primary tumor sidedness, *TBS* tumor burden score, *GMS* geometric metastatic spread model, *ARP* Aerts radiomics prior model

superior assessment of survival time in comparison to a regularized statistical analysis of TBS.

Furthermore, a clinical model incorporating important clinical baseline parameters, especially primary tumor sidedness [18, 19], yielded an inferior predictive performance than all evaluated imaging-based approaches in our study. As risk stratification of patients with mCRC is nowadays mainly based on traditional prognostic scores including clinical and pathological parameters of the primary tumor and metastases, these results underline the potential of novel imaging-based models and biomarkers for patient risk stratification.

Taking all factors into account such as stability to scan parameter variation, interpretability, and predictive performance averaged over all settings (univariable statistics, 1-year survival classification, and survival regression), the GMS appears to be the most promising and robust model. The reliable and efficient usage of models based on texture features for outcome prediction still remains a very challenging problem. The ARP model is potentially non-robust due to the susceptibility of texture measurements to technical variation [41, 42] (Table 2S of Suppl. Mat. A). This is particularly noteworthy as we used 26 different CT scanner types in 103 patients with

baseline scans due to referrals from external physicians. Although there exist approaches to calibrate texture to technical variation [43, 44], a complete absence of the influence of technical variation could fundamentally increase confidence in AI-supported systems. The GMS model showed good predictive and also statistical performance and can trivially be interpreted as a machine learning extension of TBS to integrate more fine-grained and non-linear patterns regarding metastasis distribution. This is in principle similar to the transformation of tumor heterogeneity to the radiomics setting by a machine learning–based assessment of the Aerts features. Although the effort and complexity of the WLTB GMS analysis may be higher than the assessment of the TBS, the good predictive performance, interpretability, and probable robustness to scan parameter variation could justify the effort and should be tested in larger multicenter to provide prospective evaluation as well as external validation. Notably, previous studies have largely focused on texture-based measures. The geometric metastatic spread analysis developed in our study could convert the radiology image into a "spatial map" of liver metastases. This could greatly facilitate and empower comprehensive analysis of spatial distribution, as well as its role in tumor progression and prognostication in future studies. The GMS may prove its usefulness not only in CRLM patients but could also be applied in cross-cancer studies of other gastrointestinal tumors and may be transferred as robust imaging biomarker into the setting of longitudinal studies including CT and MR imaging to assess survival prediction and treatment response. Thus, our study may provide better insights into factors associated with patient survivability by a robust data analytical model. Ultimately, holistic assessment of WLTB and robust predictive parameters such as GMS might directly translate into optimized patient management.

Limitations

Our study has a number of potential limitations. First, the study is only of medium sample size. Second, no external validation cohort was available. Another problem may arise from the variety of different CT scan protocols especially for texture quantifications of the ARP model. However, the GMS is independent from texture- or intensity-based values and is therefore expected to be invariant to technical variability.

Conclusion

Whole liver tumor burden–based measures are significantly associated with the outcome of patients with mCRC. The TBS confirms its importance for risk stratification and shows higher predictive values than clinical models or WLTB volume alone. The ARP as well as GMS model even shows a numerically higher predictive performance than TBS in a machine learning setting. The GMS as a machine learning extension of the TBS concept appears to be the most promising approach, not least due to its invariance to technical variation.

Funding information This study was funded by the German Federal Ministry of Education and Research (BMBF).

Compliance with ethical standards

Guarantor The scientific guarantor of this publication is Dr. Dominik Nörenberg.

Conflict of interest All coauthors contributed substantially to the success of our research project, which was part of a consortium and funded by an academic BMBF grant by the German Federal Ministry of Education and Research. A.M., M.S., O.T., and A.K. are employees of Siemens Healthineers and contributed with support of J.M. substantially to the statistical and machine learning analyses. However, no commercial software or product-related application has been tested in the study. The research is not financially supported by industry. Academic radiologists (D.N., T.H., S.M., M.F., E.G.), oncologists (J.H., V.H.), and doctoral students (N.J., L.L.) had full control over the data. There are no potential conflicts of interest.

Statistics and biometry Several authors have significant statistical expertise (A.M., J.M., A.K., O.T.).

Informed consent Written informed consent was obtained from all subjects (patients) in this study.

Ethical approval Institutional Review Board approval was obtained. Our retrospective study was approved by and registered with the institutional review board of the Ludwig-Maximilian University, Munich, Germany (approval no. 502-16).

Methodology
• retrospective
• prognostic study
• performed at one institution

References

1. Siegel RL, Miller KD, Jemal A (2016) Cancer statistics, 2016. CA Cancer J Clin 66(1):7–30
2. Rees M, Tekkis PP, Welsh FK, O'Rourke T, John TG (2008) Evaluation of long-term survival after hepatic resection for metastatic colorectal cancer: a multifactorial model of 929 patients. Ann Surg 247(1):125–135
3. Morris EJ, Forman D, Thomas JD et al (2010) Surgical management and outcomes of colorectal cancer liver metastases. Br J Surg 97(7):1110–1118
4. Lubner MG, Stabo N, Lubner SJ et al (2015) CT textural analysis of hepatic metastatic colorectal cancer: pre-treatment tumor heterogeneity correlates with pathology and clinical outcomes. Abdom Imaging 40(7):2331–2337
5. Beckers RCJ, Trebeschi S, Maas M et al (2018) CT texture analysis in colorectal liver metastases and the surrounding liver parenchyma and its potential as an imaging biomarker of disease aggressiveness, response and survival. Eur J Radiol 102:15–21
6. Beckers RCJ, Lambregts DMJ, Schnerr RS et al (2017) Whole liver CT texture analysis to predict the development of colorectal liver metastases-a multicentre study. Eur J Radiol 92:64–71
7. Sahu S, Schernthaner R, Ardon R et al (2017) Imaging biomarkers of tumor response in neuroendocrine liver metastases treated with transarterial chemoembolization: can enhancing tumor burden of the whole liver help predict patient survival? Radiology 283(3):883–894
8. Fleckenstein FN, Schernthaner RE, Duran R et al (2016) 3D quantitative tumour burden analysis in patients with hepatocellular carcinoma before TACE: comparing single-lesion vs. multi-lesion imaging biomarkers as predictors of patient survival. Eur Radiol 26(9):3243–3252
9. Sasaki K, Morioka D, Conci S et al (2018) The tumor burden score: a new "metro-ticket" prognostic tool for colorectal liver metastases based on tumor size and number of tumors. Ann Surg 267(1):132–141
10. Sasaki K, Margonis GA, Andreatos N et al (2017) The prognostic utility of the "tumor burden score" based on preoperative radiographic features of colorectal liver metastases. J Surg Oncol 116(4):515–523
11. De Cecco CN, Ganeshan B, Ciolina M et al (2015) Texture analysis as imaging biomarker of tumoral response to neoadjuvant chemoradiotherapy in rectal cancer patients studied with 3-T magnetic resonance. Invest Radiol 50(4):239–245
12. Dohan A, Gallix B, Guiu B et al (2019) Early evaluation using a radiomic signature of unresectable hepatic metastases to predict

outcome in patients with colorectal cancer treated with FOLFIRI and bevacizumab. Gut. https://doi.org/10.1136/gutjnl-2018-316407

13. Aerts HJ, Velazquez ER, Leijenaar RT et al (2014) Decoding tumour phenotype by noninvasive imaging using a quantitative radiomics approach. Nat Commun 5:4006

14. Bressem KK, Adams LC, Vahldiek JL et al (2020) Subregion radiomics analysis to display necrosis after hepatic microwave ablation-a proof of concept study. Invest Radiol. https://doi.org/10.1097/RLI.0000000000000653

15. Chlebus G, Schenk A, Moltz JH, van Ginneken B, Hahn HK, Meine H (2018) Automatic liver tumor segmentation in CT with fully convolutional neural networks and object-based postprocessing. Sci Rep 8(1):15497

16. van Griethuysen JJM, Fedorov A, Parmar C et al (2017) Computational radiomics system to decode the radiographic phenotype. Cancer Res 77(21):e104–e107

17. Chen Y, Chang W, Ren L et al (2020) Comprehensive evaluation of relapse risk (CERR) score for colorectal liver Metastases: Development and Validation. Oncologist. https://doi.org/10.1634/theoncologist.2019-0797

18. Kamran SC, Clark JW, Zheng H et al (2018) Primary tumor sidedness is an independent prognostic marker for survival in metastatic colorectal cancer: results from a large retrospective cohort with mutational analysis. Cancer Med. https://doi.org/10.1002/cam4.1558

19. Modest DP, Stintzing S, von Weikersthal LF et al (2017) Exploring the effect of primary tumor sidedness on therapeutic efficacy across treatment lines in patients with metastatic colorectal cancer: analysis of FIRE-3 (AIOKRK0306). Oncotarget 8(62):105749–105760

20. Ahmed S, Pahwa P, Le D et al (2018) Primary tumor location and survival in the general population with metastatic colorectal cancer. Clin Colorectal Cancer 17(2):e201–e206

21. Froelich MF, Heinemann V, Sommer WH et al (2018) CT attenuation of liver metastases before targeted therapy is a prognostic factor of overall survival in colorectal cancer patients. Results from the randomised, open-label FIRE-3/AIO KRK0306 trial. Eur Radiol 28(12):5284–5292

22. Bester L, Meteling B, Pocock N et al (2012) Radioembolization versus standard care of hepatic metastases: comparative retrospective cohort study of survival outcomes and adverse events in salvage patients. J Vasc Interv Radiol 23(1):96–105

23. Jakobs TF, Hoffmann RT, Dehm K et al (2008) Hepatic yttrium-90 radioembolization of chemotherapy-refractory colorectal cancer liver metastases. J Vasc Interv Radiol 19(8):1187–1195

24. Colloca GA, Venturino A, Guarneri D (2020) Different variables predict the outcome of patients with synchronous versus metachronous metastases of colorectal cancer. Clin Transl Oncol. https://doi.org/10.1007/s12094-019-02277-7

25. Parmar C, Leijenaar RT, Grossmann P et al (2015) Radiomic feature clusters and prognostic signatures specific for lung and head & neck cancer. Sci Rep 5:11044

26. Coroller TP, Grossmann P, Hou Y et al (2015) CT-based radiomic signature predicts distant metastasis in lung adenocarcinoma. Radiother Oncol 114(3):345–350

27. Parmar C, Rios Velazquez E, Leijenaar R et al (2014) Robust radiomics feature quantification using semiautomatic volumetric segmentation. PLoS One 9(7):e102107

28. Ding C, Peng H (2005) Minimum redundancy feature selection from microarray gene expression data. J Bioinform Comput Biol 3(2):185–205

29. Muhlberg A, Museyko O, Bousson V, Pottecher P, Laredo JD, Engelke K (2019) Three-dimensional distribution of muscle and adipose tissue of the thigh at CT: association with acute hip fracture. Radiology 290(2):426–434

30. Bousson VD, Adams J, Engelke K et al (2011) In vivo discrimination of hip fracture with quantitative computed tomography: results from the prospective European Femur Fracture Study (EFFECT). J Bone Miner Res 26(4):881–893

31. Leek JT, Scharpf RB, Bravo HC et al (2010) Tackling the widespread and critical impact of batch effects in high-throughput data. Nat Rev Genet 11(10):733–739

32. Davidson-Pilon C, Kalderstam J, Zivich P (2020) CamDavidsonPilon/lifelines: v0.23.7. Zenodo

33. Leijenaar RT, Carvalho S, Hoebers FJ et al (2015) External validation of a prognostic CT-based radiomic signature in oropharyngeal squamous cell carcinoma. Acta Oncol 54(9):1423–1429

34. Leger S, Zwanenburg A, Pilz K et al (2017) A comparative study of machine learning methods for time-to-event survival data for radiomics risk modelling. Sci Rep 7(1):13206

35. Vallieres M, Kay-Rivest E, Perrin LJ et al (2017) Radiomics strategies for risk assessment of tumour failure in head-and-neck cancer. Sci Rep 7(1):10117

36. Zhao L, Gong J, Xi Y et al (2020) MRI-based radiomics nomogram may predict the response to induction chemotherapy and survival in locally advanced nasopharyngeal carcinoma. Eur Radiol 30(1):537–546

37. Ingrisch M, Schneider MJ, Norenberg D et al (2017) Radiomic analysis reveals prognostic information in T1-weighted baseline magnetic resonance imaging in patients with glioblastoma. Invest Radiol. https://doi.org/10.1097/RLI.0000000000000349

38. Vogl TJ, Dommermuth A, Heinle B et al (2014) Colorectal cancer liver metastases: long-term survival and progression-free survival after thermal ablation using magnetic resonance-guided laser-induced interstitial thermotherapy in 594 patients: analysis of prognostic factors. Invest Radiol 49(1):48–56

39. Welch ML, McIntosh C, Haibe-Kains B et al (2019) Vulnerabilities of radiomic signature development: the need for safeguards. Radiother Oncol 130:2–9

40. Gevaert O, Mitchell LA, Achrol AS et al (2014) Glioblastoma multiforme: exploratory radiogenomic analysis by using quantitative image features. Radiology 273(1):168–174

41. Kim H, Park CM, Lee M et al (2016) Impact of reconstruction algorithms on CT radiomic features of pulmonary tumors: analysis of intra- and inter-reader variability and inter-reconstruction algorithm variability. PLoS One 11(10):e0164924

42. Ger RB, Zhou S, Chi PM et al (2018) Comprehensive investigation on controlling for CT imaging variabilities in radiomics studies. Sci Rep 8(1):13047

43. Muhlberg A, Katzmann A, Heinemann V et al (2020) The Technome - a predictive internal calibration approach for quantitative imaging biomarker research. Sci Rep 10(1):1103

44. Johnson WE, Li C, Rabinovic A (2007) Adjusting batch effects in microarray expression data using empirical Bayes methods. Biostatistics 8(1):118–127

Publisher's note Springer Nature remains neutral with regard to jurisdictional claims in published maps and institutional affiliations.

RESEARCH ARTICLE

A reproducible semi-automatic method to quantify the muscle-lipid distribution in clinical 3D CT images of the thigh

Alexander Mühlberg[1]*, Oleg Museyko[1], Jean-Denis Laredo[2], Klaus Engelke[1]

1 Institute Of Medical Physics, Friedrich-Alexander University Erlangen-Nuremberg, Erlangen, Germany,
2 AP-HP, Radiologie Ostéo-Articulaire, Hôpital Lariboisière, Université Paris VII Denis Diderot, Paris, France

* alexander.muehlberg@imp.uni-erlangen.de

Check for
updates

Abstract

Many studies use threshold-based techniques to assess *in vivo* the muscle, bone and adipose tissue distribution of the legs using computed tomography (CT) imaging. More advanced techniques divide the legs into subcutaneous adipose tissue (SAT), anatomical muscle (muscle tissue and adipocytes within the muscle border) and intra- and perimuscular adipose tissue. In addition, a so-called muscle density directly derived from the CT-values is often measured. We introduce a new integrated approach to quantify the muscle-lipid system (MLS) using quantitative CT in patients with sarcopenia or osteoporosis. The analysis targets the thigh as many CT studies of the hip do not include entire legs The framework consists of an anatomic coordinate system, allowing delineation of reproducible volumes of interest, a robust semi-automatic 3D segmentation of the fascia and a comprehensive method to quantify of the muscle and lipid distribution within the fascia. CT density-dependent features are calibrated using subject-specific internal CT values of the SAT and external CT values of an in scan calibration phantom. Robustness of the framework with respect to operator interaction, image noise and calibration was evaluated. Specifically, the impact of inter- and intra-operator reanalysis precision and addition of Gaussian noise to simulate lower radiation exposure on muscle and AT volumes, muscle density and 3D texture features quantifying MLS within the fascia, were analyzed. Existing data of 25 subjects (age: 75.6 ± 8.7) with porous and low-contrast muscle structures were included in the analysis. Intra- and inter-operator reanalysis precision errors were below 1% and mostly comparable to 1% of cohort variation of the corresponding features. Doubling the noise changed most 3D texture features by up to 15% of the cohort variation but did not affect density and volume measurements. The application of the novel technique is easy with acceptable processing time. It can thus be employed for a comprehensive quantification of the muscle-lipid system enabling radiomics approaches to musculoskeletal disorders.

OPEN ACCESS

Citation: Mühlberg A, Museyko O, Laredo J-D, Engelke K (2017) A reproducible semi-automatic method to quantify the muscle-lipid distribution in clinical 3D CT images of the thigh. PLoS ONE 12 (4): e0175174. https://doi.org/10.1371/journal.pone.0175174

Editor: Cristina Óvilo, INIA, SPAIN

Received: October 3, 2016

Accepted: March 21, 2017

Published: April 28, 2017

Data Availability Statement: All relevant data is within the paper.

Funding: The work was supported by the FORMOSA project (AZ-1044-12) of the Bavarian Research Foundation, Munich, Germany.

Competing interests: The authors have declared that no competing interests exist.

Introduction

Loss of muscle function, as a result of diseases such as neuropathies and myopathies on one hand, or sarcopenia, the age-related loss of muscle mass and function on the other, results in reduced function. Muscle diseases also play a role in the pathogenesis of osteoporosis, falling incidence and bone frailty and fractures. Muscle biopsy is the gold standard for muscle assessment but is invasive and only evaluates a small sample, not always representative, of the relevant muscle tissue. Two different approaches are currently available for *in vivo* 3D muscle and fat evaluation. The first consists of a qualitative or, at best, semi-quantitative grading of muscle structures based on a washed-out and moth-eaten muscle appearance in MRI [1–3] or CT [4, 5] images. The second approach consists of a quantitative measurement of muscle volume or cross-sectional area (CSA) in both MRI or CT images and density measurements with CT [6–10]

While MRI has advantages in the qualitative assessment of muscle structures due to its superior soft tissue contrast, quantitative MRI analyses are affected by complex artifacts such as proton spin inhomogeneities, which usually depend on the specific MR scanner and acquisition sequence. In contrast, CT is less affected by technical variations [11], and provides a higher spatial resolution, which is an important advantage for fine-grained measurements as 3D texture. In addition, muscle density cannot be measured by MRI.

For a radiomics [12] approach to the muscle-lipid system (MLS), a reproducible, robust and fast segmentation and quantification method is required, to exploit more diagnostic relevant information from CT images. This study specifically targeted the thigh instead of the whole upper or even combined upper and lower legs, because many CT scans originally performed to determine bone mineral density of the hip extend to the upper to mid shaft of the femur.

Tissue composition of the thigh

For a better understanding of the specific segmentation and quantification techniques described in the following sections, a brief overview of the various tissues and compartments of the thigh will be given. The outer layer of the leg is the subcutaneous adipose tissue (SAT) with the skin as an outer surface and the deep fascia (F) as an inner surface (Fig 1). The volume of interest (VOI) inside the fascia (VOI$_{IF}$: IF = Intrafascia) can be separated into the femoral bone, anatomical muscles (M) consisting of muscle tissue and adipocytes within the muscle border and perimuscular adipose tissue (PAT) separating the anatomical muscles. In the present work, no attempt was made to separate individual muscles. Therefore, M will always refer to the combination of all muscles. Adipose tissue (SAT and PAT) consists of adipocytes (fat cells) containing lipids.

Muscle tissue can be further differentiated into myocytes (muscle fibers), which may contain intramyocellular (IML) [13], and extramyocellular lipids (EML). In a CT image, IML can be measured indirectly only via the CT muscle density, which will decrease with increasing IML [14]. EML are adipocytes embedded within M among the muscle fibers. EML is typically present as adipocyte clusters, which, depending on their size and the spatial resolution of the CT image, can be directly segmented because their CT value differs from that of the surrounding muscle tissue. According to reference [15] IMAT (intermuscular AT) will include both PAT and EML in the present study (Fig 1), although some authors [16] refer to PAT only.

Materials and methods

The integrated approach can be divided into four steps: (A) definition of a global VOI analysis of the thigh, (B) segmentation of the fascia to delineate SAT and IMAT, (C) segmentation of

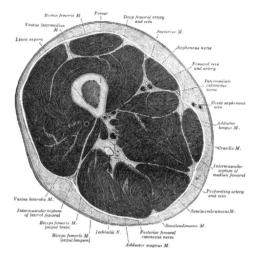

Fig 1. Muscle-lipid system (MLS) of the thigh. Cross section of the femur (left) and schematic composition (right) with muscle (dark grey) and adipose tissue (light grey). The fascia is shown in yellow and the periosteal bone surface in green. AT within the fascia is termed intermuscular AT (IMAT), outside subcutaneous adipose tissue (SAT). IMAT within the anatomical muscle M (magenta contours define the border of the Semitendinosus) is termed extramyocellular lipids (EML) and outside perimuscular AT (PAT). In CT images intramyocellular lipids (IML) can only be measured indirectly by a lower average CT value compared to that of pure muscle tissue.

https://doi.org/10.1371/journal.pone.0175174.g001

muscle and lipid VOIs within VOI_{IF} and (D) definition of features characterizing the muscle-lipid distribution within VOI_{IF}. CT scans of the proximal femur, obtained earlier [17] to measure bone mineral density in subjects with high fracture risk, were used. Subjects were scanned on top of a calibration phantom to calculate BMD from the measured CT values. The femoral bone was segmented as part of a previous study [18].

A. Analysis VOI

The thigh surface was first segmented using a threshold-based volume growing. The threshold, which was empirically selected as 70%-threshold between highest and lowest CT values, was determined from the CT value spectrum of a sphere automatically positioned on the thigh surface.

The muscle-lipid analysis was carried out at the level of the upper femoral shaft (VOI_{US}; Fig 2). Segmentation of the VOI_{US} started with the automatic determination of two points: the first was the center of mass of the femur in plane A, which was perpendicular to the femoral shaft axis and intersected the mid-height of the lesser trochanter. The second point was the voxel of the femoral head at a maximal distance from A. The projected distance between the two points onto the scanner z-axis was used as an anatomic size- and pose-specific distance d. VOI_{US} consisted of n = 0.5 d / s (s: CT slice thickness) slices. The most proximal slice was defined as the distal end of the acetabulum.

B. Segmentation of the fascia

Accurate fascia segmentation is essential for the separation of SAT from muscle and IMAT. However, fascia segmentation is difficult since it is a very thin structure with low contrast

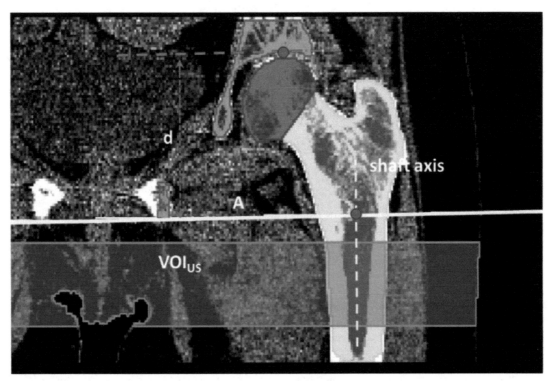

Fig 2. Definition of the analysis VOI. VOI_{US} defined by 1: The center of mass of the femur in plane A, which is perpendicular to the shaft axis and intersects the center of the trochanter minor and 2.: The voxel of the femoral head with maximum distance from A. The projected distance between the two points onto the scanner z-axis is used as an anatomic size- and pose-specific distance d. VOI_{US} contains n = 0.5 d / s (s: slice thickness) slices; the most proximal slice contained the distal end of the acetabulum (not visible here).

https://doi.org/10.1371/journal.pone.0175174.g002

relative to the surrounding AT. Pure muscle tissue has higher CT values than AT. However, muscle tissue with high lipid infiltration has CT values close to those of AT. Muscle had also to be distinguished from blood vessels, edema, dermis and genital organs through the following steps:

1. The first step was a gross identification of muscle, based on its CT appearance: a grade G in the range [1, 3] was first set by the operator where G = 3 denotes moth-eaten and washed-out muscle structures (Fig 3a and 3b). In addition, a contrast value (C), relative to AT and water was assigned to each voxel (v) according to a linear scale where C = 0 for a CT value equal to the average CT value of the adipose tissue (CT_{AT}) and C = 1 for a CT value equal to that of water. CT_{AT} was determined in a VOI_{AT} resulting from simple volume growing using a CT value range [-190 HU; -30 HU], which started within the sphere on the thigh surface. CT_{H2O} was determined from the in-scan calibration phantom.

In order to classify a voxel v as potential muscle, four empirical conditions had to be fulfilled: (1) a minimum distance of 10 voxels from the body surface; (2) connection with the femur; (3) a contrast C above the minimum contrast C_{min} characterizing the washed-out appearance; and (4) an aggregation α above the minimum aggregation α_{min} characterizing the

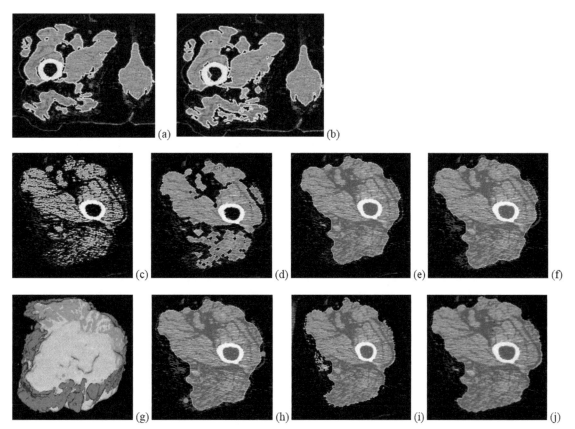

Fig 3. Multi step 3D hierarchical segmentation of the volume of interest inside the fascia (VOI$_{IF}$) exemplified in one 2D axial slice.
Step 1: Connected components of right leg (VOI$_{RC}$) for G = 1 (a) and G = 3 (b) (G: Grade of muscle appearance, see text). Step 2: Probable muscle voxels within processed left leg (VOI$_{LC}$) (c) and compactification resulting in VOI$_{IFA}$ (d and e) (IFA: Approximation of IF). Step 3: refinement of VOI$_{IFA}$. Candidate voxels close to FA that may belong to VOI$_{IF}$ (f and g), seed points to select clusters that belong to VOI$_{IF}$ (h), clusters that were not selected stay porous (i) and are removed by morphological smoothing yielding the final result (j). User-dependence is indicated by blue color.

https://doi.org/10.1371/journal.pone.0175174.g003

moth-eaten appearance, where:

$$\alpha(v) = \frac{\sum_{v'=1}^{26} \theta(C(v') - C_{min})}{26}$$

$$C_{min} = 0.75 - 0.15\,G$$

$$\alpha_{min} = 0.85 - 0.05\,G$$

Θ denotes the Heaviside function and v' is a voxel in the 26-neighborhood of the voxel v. The aggregation value is the ratio of neighboring voxels with the contrast above C_{min}.

From the resulting voxel clusters only the two largest connected components within VOI$_{US}$ were kept. According to the position of their center of mass relative to the femur, the two components were classified as right (VOI$_{RC}$) or left (VOI$_{LC}$). VOI$_{RC}$ and VOI$_{LC}$ were independently processed by simple morphological closing using a z-axis elongated elliptical structure

element to account for the cylindrical shape of the fascia and the predominant extension of the muscles along the leg axis, and contour-filling.

2. The processed VOI_{RC} and VOI_{LC} exclude the dermis, connective tissue and genital organs, whose CT values are typically similar to those of muscle. Within each of these VOIs, a simple contrast value threshold was used to identify probable muscle voxels (Fig 3c). These voxels were compacted (Fig 3d and 3e) resulting in a structure named VOI_{IFA}, an approximation of the intrafascia VOI. Its surface was named fascia approximation (FA).

3. FA was already a very good approximation of the fascia when the amount of IMAT was low, i.e. in younger healthy subjects, in which the fascia is usually in direct contact with the muscle. However, with increasing adipose tissue infiltration, the amount of IMAT increases and the fascia is often bordered by adipose tissue on either side, SAT on the outer and IMAT on the inner side. In many areas, the CT contrast of the fascia is very low, preventing the direct detection of a closed surface. Instead, in our study FA was used as a start point to select additional voxels, which were likely to belong to VOI_{IF}. Specifically, voxels, which were connected to FA, were not connected with the dermis and were located at a maximal distance of 15 voxels from FA, were identified as candidates (Fig 3f and 3g). Connection to the dermis was defined as voxels with $C > 1$ connected with the thigh surface. Based on volume growing, connection to FA was ensured by a local noise adaptive threshold. Namely, a voxel v was accepted by the volume growing if f was true:

$$f = \begin{cases} \text{true if } C > 1 \\ g \text{ if } 1 \geq C \geq 0.5 \\ \text{false else} \end{cases}$$

where g depended on the mean CT value (mean), and the standard deviation (SD) in the 26-neighborhood of v:

$$g = \begin{cases} \text{true if } CTvalue \geq mean_{26} - 2SD_{26} \\ \text{false else} \end{cases}$$

The 2SD factor accounted for noise in the CT data. Next, starting from a seed point set by the operator in the space between FA and a local group of the new candidate voxels, 'rays' were sent out isotropically (26 directions). If 65% or more of these rays were reflected by one of the candidate voxels or by FA, the status of the voxel was changed to 'shielded' [18] (Fig 4). All shielded and candidate voxels were merged with VOI_{IFA} (Fig 3i), which was then filtered for porous or non-compact structures by a morphological smoothing. The resulting VOI was called intrafascia (VOI_{IF}, Fig 3j), its surface defined the 3D fascia. As shown in Fig 3h, often multiple seed points were required to process the complete dataset. The operator could also select not to add certain conglomerates of candidate voxels to VOI_{IF}.

4. Optionally, a fascia-specific modification of the explicit deformable model by Mastmeyer [19] could be utilized to attract the fascia to the nearest maximum under constraint of a given regularity of the 3D surface. This step was useful, if neither strong edema nor very washed-out muscle structures were present in the image, which was typically the case for younger patients.

C. Muscle and lipid VOIs

In order to facilitate a comprehensive analysis of the relation between muscle and lipids, various VOIs were determined. With the exception of VOI_{SAT}, all of them were located within VOI_{IF}. Specifically an anatomical muscle VOI_M and a muscle tissue VOI_{MT} were defined.

							░		░						░
	░		-	-	░	-	x	x	x	░					
		-	-	x	x	x	x	h	x	x	░				
░		-	-	x	x	x	x	x	x	x	x	x	-	░	
	-	-	-	-	-	x	░	h	h	h	h	x	-		
	-	x	-	-	x	x				x	x	x	-		
	░	x	x	x							x	-			
░	x	x	x								-		░		
	x	h	h								-	-	-		
	x	h	h								-	x			
	x	x	x	x							x	x	x		
	x	x	x	x	h	x			x	x	x	x	x		
	░	x	x	h	h	x	x	x	h	x	x	x	-	-	-
░	x	x	x	x	x	x	x	x	x	x	x		-	░	
	x	x	x	x	x	░	x	x	x	░	-				
-	-	-	░	-	-	-	x	x	x	-					
				-	-	░	-	-							

Fig 4. 3D ray reflection model. Approximation of the intrafascia VOI (yellow), candidate voxels (purple) and seed point (blue) for the ray reflection model: The space between IFA and candidate voxels that were connected with the seed were either x: shielded, -: not-shielded or h: filled by contour-filling.

https://doi.org/10.1371/journal.pone.0175174.g004

1. VOI_{SAT} was obtained by subtracting VOI_{IF} and all voxels connected with the dermis from VOI_{US}. The spectrum of CT values of VOI_{SAT} (Fig 5) was used to define an AT threshold (T_{AT}), which differed from CT_{AT} used for the fascia segmentation. Typically, SAT is very homogenous but may contain edema and blood vessels that have higher CT values than

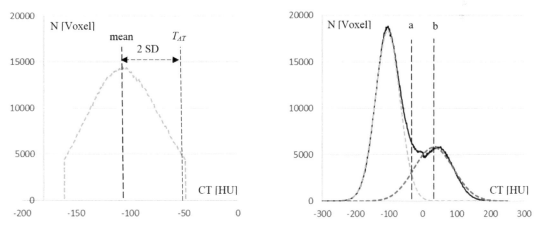

Fig 5. Image-specific definition of thresholds for anatomical muscle and muscle tissue. Left graph: definition of the adipose tissue threshold T_{AT} from SAT*; SD: standard deviation. Right graph: CT value spectrum of the combined subcutaneous adipose tissue and muscle tissue VOIs (VOI_{MLS}) (black). Gaussian mixture model used to fit adipose (yellow) and muscle tissue (magenta) distributions; a: intersection between the two curves, b: peak of the muscle tissue curve.

https://doi.org/10.1371/journal.pone.0175174.g005

adipose tissue. Therefore high and low CT values of the spectrum with frequencies lower than 30% of the most frequent CT value were cut. In order to determine T_{AT} from a symmetric spectrum to prevent an influence of noise on the mean value, the lower tail of the spectrum was also cut resulting in a spectrum SAT*. T_{AT} was defined as mean + 2SD from the mean and the standard deviation of the SAT* CT value histogram (Fig 5 left).

2. VOI_M was determined by subtracting VOI_{PAT} from VOI_{IF} under the assumption that PAT but not EML was connected with SAT. VOI_{PAT} was defined by connection with the fascia under the condition that the CT value of the voxel under consideration was smaller than T_{AT}.

3. For the definition of VOI_{MT}, a Gaussian mixture model (GMM) was employed in combination with a Levenberg-Marquard optimization algorithm [20] to fit two Gaussian curves to the CT value spectrum of the combined SAT and IF VOIs ($VOI_{SAT} \cup VOI_{IF} = VOI_{MLS}$) (Fig 5, right). One fit curve represented AT and one MT. For AT, height, peak and width of the SAT* distribution were used for initialization of the fit procedure. For MT the ratio of the number of voxels of VOI_{AT} versus those of VOI_{MLS} was used to initialize the height. A phantom based CT value was used to initialize its peak. VOI_{MT} was then defined by 3D volume growing inside VOI_{IF} starting from seed points defined as voxels with CT values higher than the peak b of the fitted MT curve. Voxels were included if their CT values were higher than that of the intersection a between AT and MT distributions (Fig 5, right).

4. Finally, in order to exploit the fact that the CT values of MT also reflect IML, similar to [21] an abstract description for the muscle-lipid distribution was formulated. For each voxel of VOI_{IF}, a muscle concentration between 0% and 100% was determined using T_{AT} and T_{HDM} with T_{AT} defining a muscle concentration of 0% and T_{HDM}, a muscle concentration of 100% (HDM: high density muscle). It was determined as mean CT value minus the SD determined in a sphere placed in the muscle tissue of young climbers. Results from 30 subjects were averaged and resulted in a CT value of 35 HU. In order to account for calibration differences among scanners, T_{HDM} was finally defined as 35 HU + CT_{H2O}. Thus, apart from the constant of 35HU, it only depended on the water value of the in-scan calibration phantom. The two CT value thresholds T_{HDM} and T_{AT} were used to define 6 bins with decreasing "muscle concentrations". Bin 6 (B6) defined HDM for $\geq$ 100% muscle concentration, B5 muscle concentration for 75–99%, B4 for 50–74%, B3 for 25–49%, B2 for 0–24% and intermuscular adipose tissue (IMAT) for $\leq$ 0%.

D. Feature extraction

Numerous image features can be calculated in the segmented VOIs to describe the muscle-lipid distribution. Only few features quantifying volume, density and structure will be discussed in the present report to investigate the reproducibility and stability of the segmentation and the feature extraction.

Densities were determined either from the segmentation-based or from GMM-based procedures. Segmentation-based densities were defined as average CT values in a VOI after subtraction of CT_{H2O}. GMM-based densities were defined as peaks of IF or MLS distribution curves again after subtraction of CT_{H2O} i.e. $D_{IF}^{GMM}(MT)$ or $D_{MLS}^{GMM}(MT)$. rV defined a volume of a VOI relative to VOI_{MLS} or VOI_{IF}.

A more advanced 3D descriptor was the average grain size G_{avg} as determined by granulometry [22]. Porous and atrophic MT and HDM should have a smaller G_{avg}. Fractal dimension (FD) [23] was used as a measure of 3D texture roughness. The 3D texture of dystrophic muscles should have a rougher appearance. Further, sphericity Ψ of M and MT was quantified. Ψ should be smaller for irregular surfaces and elongated muscle shape in case of pathologies.

E. Validation

Validation of the segmentation stability and features included determination of the intra- and inter-operator precision errors and the impact of noise and of CT_{H2O} derived from the in-scan calibration phantom.

25 datasets from elderly subjects (age: 75.6 ± 8.7) of the EFFECT study [30] with a washed-out and moth-eaten muscle appearance and high levels of IMAT and edema were processed. In these subjects, FA differed largely in most cases from the true fascia and an operator-inter-action was needed in such cases. For the precision analysis, all 25 datasets were analyzed three times on different days by the same operator for intra-operator-precision and once by three different operators after an initial training for inter-operator-precision. Precision errors were calculated as CV_{RMS} values [24]. In order to assess the diagnostic value of a feature, the CV_{RMS} was compared to the biological variation of the 25 subjects measured as percentage coefficient of variation divided by a factor of 100 (CV_{100}).

Gaussian noise was added to the CT datasets acquired at 120 kV with 170 mAs to simulate a lower exposure of 135 mAs. Finally, CT_{H2O} obtained from the in-scan calibration phantom was changed by ± 5 HU.

Results

Fig 6 shows segmentation results for three different patients, in which muscles were heavily infiltrated by adipose tissue and fascia not always in direct contact with the muscle. Thus, segmentation required operator interactions. However, even in these challenging cases the hierarchical segmentation process limited the required operator interaction to a few clicks by setting some seed points as shown in Fig 4f.

Depending on the segmentation complexity the complete processing time per CT dataset was around 12 min (i5 processor 5GHz, 4GB RAM). Table 1 shows results for intra- and inter-operator analysis precision errors. There were no repeat scans of patients. For comparison, Table 1 also shows the CV_{100} of the same feature from the 25 subjects. All CV_{RMS} results were below 2% and even below 0.6% in most cases reflecting the low impact of the operator interactions. Most of the CV_{RMS} results were comparable to the CV_{100} value.

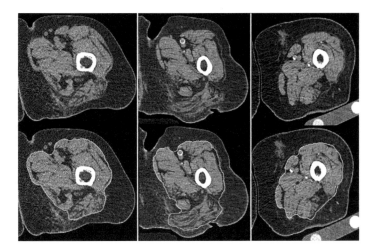

Fig 6. Segmentation results for three subjects with washed-out and moth-eaten muscles and strong edema. First row: native images; second row: segmented fascia (yellow).

https://doi.org/10.1371/journal.pone.0175174.g006

Table 1. Reanalysis precision errors as CV$_{rms}$ in per cent and biological variance as CV/100 in percent for density, relative volume and features for different VOIs. In each row intra/interoperator CV$_{rms}$ results are on top and CV$_{100}$ results below. D: density; D$_{If}^{GMM}$, D$_{MLS}^{GMM}$: density derived from Gaussian mixture model applied to of VOI$_{IF}$ or VOI$_{MLS}$, respectively; rV$_{MLS}$, rV$_{IF}$: Volume relative to VOI$_{IF}$ or VOI$_{MLS}$, respectively; G$_{avg}$: average grain size; FD: fractal dimension; Ψ: sphericity. VOIs: SAT: subcutaneous adipose tissue; IF: intrafascia; M: muscle; MT: muscle tissue; HDM: high density muscle; IMAT: intermuscular adipose tissue. Note: not all features are determined for all VOIs.

	SAT	IF	M	MT	HDM	IMAT
D	0.06/0.10/	0.53 / 1.05	0.21 / 0.27	0.19 / 0.20	0.03 / 0.10	0.06 / 0.15
	0.07	0.79	0.27	0.23	0.08	0.18
D$_{IF}^{GMM}$				1.60 / 1.90		
				0.18		
D$_{MLS}^{GMM}$				0.53 / 0.56		
				0.44		
rV$_{IF}$			0.11 / 0.24	0.10 / 0.27	0.20 / 0.33	0.51 / 1.90
			0.27	0.07	0.22	0.55
rV$_{MLS}$	0.15 / 0.26	0.25 / 0.40				
	0.12	0.15				
G$_{avg}$				0.24 / 1.1	0.04 / 0.15	0.18 / 0.31
				0.27	0.26	0.16
FD		0.03 / 0.05	0.04 / 0.05	0.05 / 0.06		
		0.017	0.019	0.021		
Ψ			0.57 / 0.62	0.55 / 0.70		
			0.20	0.23		

https://doi.org/10.1371/journal.pone.0175174.t001

Table 2 shows the percentage changes after simulating an exposure level of 135 mAs, which would double the noise when compared to the 170 mAs exposure, which was actually used for the CT acquisition. Density and volume measurements of VOI$_{SAT}$ and VOI$_{IF}$ were not affected by higher noise. For most other features, the noise related change was about one magnitude higher than the reanalysis precision error at 170 mAs. Highest noise related changes were measured for the volume of VOI$_{IMAT}$ relative to the volume of VOI$_{IF}$ and for the average grain size of VOI$_{MT}$

For the features or segmentations underneath the fascia that were calibrated by water calibration or more specifically CT$_{H2O}$, minor variations (± 5 HU) in the CT$_{H2O}$ value, which in our case was obtained from the in-scan calibration phantom, caused a much larger effect than a increase in noise. Apart from the density of HDM, a small change (± 5 HU) in the CT$_{H2O}$

Table 2. Percentage changes for density, relative volume and features for different VOIs when an exposure level of 135 mAs was simulated. D: density; D$_{IF}^{GMM}$, D$_{MLS}^{GMM}$: density derived from Gaussian mixture model applied to of VOI$_{IF}$ or VOI$_{MLS}$, respectively; rV$_{MLS}$, rV$_{IF}$: Volume relative to VOI$_{IF}$ or VOI$_{MLS}$, respectively; G$_{avg}$: average grain size; FD: fractal dimension; Ψ: sphericity. VOIs: SAT: subcutaneous adipose tissue; IF: intrafascia; M: muscle; MT: muscle tissue; HDM: high density muscle; IMAT: intermuscular adipose tissue.

	SAT	IF	M	MT	HDM	IMAT
D	0.0	0.0	5.1	3.5	3.4	2.6
D$_{IF}^{GMM}$				4.0		
D$_{MLS}^{GMM}$				4.2		
rV$_{MLS}$	0.0	0.0				
rV$_{IF}$			2.0	1.1	0.71	15.0
G$_{avg}$				14.1	4.9	2.1
FD		0.89	0.79	0.99		
Ψ			3.8	8.1		

https://doi.org/10.1371/journal.pone.0175174.t002

Table 3. Effect of water calibraton. A simulated addition / subtraction of 5 HU from CT_{H2O}.caused the percent change shown in the table. D: density; D_{IF}^{GMM}, D_{MLS}^{GMM}: density derived from Gaussian mixture model applied to of VOI_{IF} or VOI_{MLS}, respectively; SAT: subcutaneous adipose tissue; IF: intrafascia; M: muscle; MT: muscle tissue; HDM: high density muscle muscle.

	IF	M	MT	HDM
D	27.4 / 27.4	15.8 / 15.8	13.4 / 13.4	3.2 / 3.5
D_{IF}^{GMM}			11.6 / 11.6	
D_{MLS}^{GMM}			17.8 / 17.8	

https://doi.org/10.1371/journal.pone.0175174.t003

value caused density differences of more than 10% (Table 3) although relative to CV_{100} results were comparable.

Discussion

To our knowledge, this is the first study presenting a hierarchical 3D approach to segment and quantify the different muscular and lipid components of the thigh. One critical step is segmentation of the fascia, which separates SAT from IMAT and muscle from edema, blood vessels, genital organs etc. Calibration based on the relatively homogeneous subcutaneous adipose tissue and the CT value of water, which, in the present work, was determined using an in-scan calibration phantom, is another important characteristic of our method. The third critical step is the attempt to characterize muscle in different manners: as anatomical muscle, as muscle tissue, and in an abstract way—comparable to grey matter concentration in neuroimaging [25]—as muscle concentrations.

The most difficult step was the segmentation of the fascia covering the muscles, which, in elderly subjects, is no longer in tight contact with the muscle surface. The main purpose of most of the preparatory steps illustrated in Fig 3a–3i was the identification of a search space for voxels likely to belong to VOI_{IF}. A second purpose was the differentiation of external genital organs, vessels, dermis and edema from the moth-eaten and washed out muscle as shown in Fig 3. CT value differences, for example between edema and washed out muscle structures, are small and therefore techniques based on a simple global threshold to separate muscle and adipose tissue will have severe limitations. Our method based on fascia segmentation and integration of anatomical knowledge does not have such inconvenients (Fig 6).

A limitation to our study is the lack of comparison to a gold standard such as manual segmentation. However, manual segmentation is highly fastidious and therefore almost not feasible in 3D and not used by radiologists who prefer to perform semi-quantitative grading based on gross muscle CT appearance [1–3]. We also have not applied our method to patients with diseases that largely destroy muscle tissue. In this case, the separation of adipose and muscle tissue as shown in Fig 5 may fail. Further, we have not analyzed any longitudinal data in order to quantify the effect of age, disease or treatment related morphological changes on segmentation. Finally, we have not analyzed the impact of spatial resolution on segmentation. CT images used here had a slice thickness of 1 mm and an in plane voxel size of $(0.8 \text{ mm})^2$ and were reconstructed with a medium kernel. This provides an adequate balance between spatial resolution, noise and radiation exposure. Different settings of CT acquisition and reconstruction parameters would be required to evaluate the impact of spatial resolution.

The very low intra- and inter-operator precision errors of our method, below 1% (Table 1) is due to the high degree of automation of the analysis. Manual interactions as those shown in (Fig 3h) and pre-grading of muscle structures (Fig 3a and 3b) had minimal impact on precision. However, we did not carry out a precision study including patient repositioning, which

may increase precision error, since muscle shape may change across multiple scans. Repeating CT acquisitions in humans is difficult due to ethical issues.

Precision error has to be put in perspective with the changes to be measured, either longitudinally to assess changes in a given subject or cross-sectionally to compare different subjects. Table 1 shows that the reanalysis precision errors were comparable with 1% of the variation of the corresponding measurement in 25 elderly female subjects, again a very good result.

It remains speculative whether other segmentation methods [26–29] would improve the results obtained here. When applying a hierarchical approach, selection of the different steps and their order remains subjective. To our knowledge, starting with a segmentation of the fascia, an anatomical boundary at the superficial aspect of the muscles is an innovative method. However, a comparison of the different segmentation methods remains to be done.

Existing techniques for muscle measurements on CT images are restricted to measure muscle volume or area and density. Several authors used fixed CT value thresholds for muscle segmentation [30–32]. Use of fixed thresholds to carry out muscle segmentation does not address differences in muscle density caused by variable degrees of muscle lipid infiltration, which vary widely among muscles.

Use of fixed CT values or Hounsfield units also does not address variation with scanner type and manufacturer. Although all clinical CT scanners are routinely calibrated to water, deviations of ±10 HU are frequent. In addition, a second value is required for calibration of muscle and adipose tissue. In the field of osteodensitometry [33], specific in-scan phantoms are used for the calibration of CT to measure bone mineral density.

In the method developed here, calibrated CT values were used for two different purposes. In the segmentation process CT_{H2O}, the CT value of the water insert of the calibration phantom and the mean CT value of adipose tissue were used to define a contrast scale, which was primarily used to define probable muscle voxels and to define upper and lower limits for the local adaptive threshold based ray reflection model. However, the final fascia resulted from the combination of several advanced image-processing procedures rather than from simple thresholding.

For segmentation of VOIs inside the fascia, two CT values derived from image information were used, CT_{H2O} and a second CT value defined precisely using the SAT CT value histogram (Fig 6). SAT is one of the most homogenous adipose tissues in the human body [34], and is well-suited for this purpose but, in order to exclude higher density voxels due to vessels and edema contained into the SAT, the histogram was trimmed, which resulted in a unique subject-specific threshold T_{AT}. Finally, the threshold for 100% muscle depended only on CT_{H2O} and a constant determined from muscle tissue of a group of young athletes.

Thus, our segmentation method uses a calibration that depends only on subject-specific internal CT values and CT_{H2O} and is therefore independent from the scanner model and manufacturer. An accurate determination of CT_{H2O} is critical: as shown in Table 3, a change of ± 5 HU in CT_{H2O} results in a change in density values of up to 30%, emphasizing that deviations from the regular water calibration of clinical CT scanners must be controlled.

Another critical characteristic of segmentation and feature extraction is the sensitivity to noise. To test our method, we simulated a 100% noise increase by retrospectively adding Gaussian noise to the CT images. Table 2 shows that density and volume measurements of SAT were not affected by a 100% noise increase, which is very important with regard to the calibration. With the exception of rV_{IF} of V_{IMAT}, effects on muscle density and volume measurements were below 5%. Not surprisingly, an increase in noise had larger effects on structural measurements like 3D texture features. Automatic exposure control techniques, nowadays often used in clinical CT [35], can approximately guarantee similar noise levels between patients. A new data-driven method to identify and reduce the impact of technical variation—

such as noise—on features is currently developed in our group for e.g. retrospective analyses of studies without automatic exposure control techniques.

Conclusion

We have developed an innovative semi-automatic approach for segmentation and quantification of the muscle-lipid distribution in CT images of the thigh. An important characteristic is the calibration combining subject-specific internal subcutaneous adipose tissue with an externally derived water calibration. Another characteristic is the hierarchical segmentation of the fascia consisting of a pre-grading based on established diagnostic criteria and a local adaptive threshold-based 3D ray reflection model which requires only minimal user interaction and offers an automatic refinement by an explicit deformable model. Extracted features included density- and volume-based muscle and lipid measurements as well as advanced 3D features for a detailed quantification of the muscle-lipid distribution underneath the fascia. The precision of the method was excellent and processing speed acceptable, enabling a comprehensive radiomics approach to musculoskeletal lipid distribution disorders from standard-of-care CT images. Results reported here concern the thigh but the same concept may be applicable to other body parts although some anatomy specific changes will be required in the segmentation step.

Acknowledgments

The authors acknowledge support from the Bavarian Research Foundation (FORMOSA project, grant number AZ-1044-12). The present work was performed in partial fulfilment of the requirements for obtaining the PhD degree Dr. rer. biol. hum. at the University of Erlangen-Nuremberg. The authors would like to thank Bastian Gerner and Andreas Friedberger for their repeat segmentation of data sets required for the inter-operator-precision analysis.

Author Contributions

Conceptualization: AM.

Data curation: AM JL.

Formal analysis: AM.

Funding acquisition: KE.

Investigation: AM.

Methodology: AM.

Project administration: KE.

Resources: AM JL KE.

Software: AM OM.

Supervision: KE.

Validation: AM.

Visualization: AM.

Writing – original draft: AM KE JL.

Writing – review & editing: KE AM OM JL.

References

1. Kornblum C, Lutterbey G, Bogdanow M, Kesper K, Schild H, Schröder R, et al. Distinct neuromuscular phenotypes in myotonic dystrophy types 1 and 2. Journal of neurology. 2006; 253(6):753–61. https://doi.org/10.1007/s00415-006-0111-5 PMID: 16511650

2. Fischer D, Kley R, Strach K, Meyer C, Sommer T, Eger K, et al. Distinct muscle imaging patterns in myofibrillar myopathies. Neurology. 2008; 71(10):758–65. https://doi.org/10.1212/01.wnl.0000324927.28817.9b PMID: 18765652

3. Mercuri E, Talim B, Moghadaszadeh B, Petit N, Brockington M, Counsell S, et al. Clinical and imaging findings in six cases of congenital muscular dystrophy with rigid spine syndrome linked to chromosome 1p (RSMD1). Neuromuscular Disorders. 2002; 12(7):631–8.

4. Swash M, Brown MM, Thakkar C. CT muscle imaging and the clinical assessment of neuromuscular disease. Muscle & nerve. 1995; 18(7):708–14.

5. Wattjes MP, Fischer D. Neuromuscular imaging: Springer; 2013.

6. Goodpaster BH, Thaete FL, Kelley DE. Thigh adipose tissue distribution is associated with insulin resistance in obesity and in type 2 diabetes mellitus. The American journal of clinical nutrition. 2000; 71 (4):885–92. PMID: 10731493

7. Goodpaster BH, Park SW, Harris TB, Kritchevsky SB, Nevitt M, Schwartz AV, et al. The loss of skeletal muscle strength, mass, and quality in older adults: the health, aging and body composition study. The Journals of Gerontology Series A: Biological Sciences and Medical Sciences. 2006; 61(10):1059–64.

8. Goodpaster BH, Carlson CL, Visser M, Kelley DE, Scherzinger A, Harris TB, et al. Attenuation of skeletal muscle and strength in the elderly: The Health ABC Study. Journal of Applied Physiology. 2001; 90 (6):2157–65. PMID: 11356778

9. Lang T, Cauley JA, Tylavsky F, Bauer D, Cummings S, Harris TB. Computed tomographic measurements of thigh muscle cross-sectional area and attenuation coefficient predict hip fracture: The health, aging, and body composition study. Journal of Bone and Mineral Research. 2010; 25(3):513–9. https://doi.org/10.1359/jbmr.090807 PMID: 20422623

10. Snijder M, Visser M, Dekker J, Goodpaster B, Harris T, Kritchevsky S, et al. Low subcutaneous thigh fat is a risk factor for unfavourable glucose and lipid levels, independently of high abdominal fat. The Health ABC Study. Diabetologia. 2005; 48(2):301–8. https://doi.org/10.1007/s00125-004-1637-7 PMID: 15660262

11. Kumar V, Gu Y, Basu S, Berglund A, Eschrich SA, Schabath MB, et al. Radiomics: the process and the challenges. Magnetic resonance imaging. 2012; 30(9):1234–48. https://doi.org/10.1016/j.mri.2012.06.010 PMID: 22898692

12. Gillies RJ, Kinahan PE, Hricak H. Radiomics: images are more than pictures, they are data. Radiology. 2015; 278(2):563–77. https://doi.org/10.1148/radiol.2015151169 PMID: 26579733

13. Coen PM, Goodpaster BH. Role of intramyocelluar lipids in human health. Trends in Endocrinology & Metabolism. 2012; 23(8):391–8.

14. Goodpaster BH, Kelley DE, Thaete FL, He J, Ross R. Skeletal muscle attenuation determined by computed tomography is associated with skeletal muscle lipid content. Journal of Applied Physiology. 2000; 89(1):104–10. PMID: 10904041

15. Marcus RL, LaStayo PC, Ryana AS, Odessa Addison D. Intermuscular Fat: A Review of the Consequences and Causes.

16. Miller CG, Krasnow J, Schwartz LH. Medical imaging in clinical trials: Springer; 2014.

17. Bousson VD, Adams J, Engelke K, Aout M, Cohen-Solal M, Bergot C, et al. In vivo discrimination of hip fracture with quantitative computed tomography: results from the prospective European Femur Fracture Study (EFFECT). Journal of Bone and Mineral Research. 2011; 26(4):881–93. https://doi.org/10.1002/jbmr.270 PMID: 20939025

18. Kang Y, Engelke K, Kalender WA. A new accurate and precise 3-D segmentation method for skeletal structures in volumetric CT data. IEEE transactions on medical imaging. 2003; 22(5):586–98. https://doi.org/10.1109/TMI.2003.812265 PMID: 12846428

19. Mastmeyer A, Engelke K, Fuchs C, Kalender WA. A hierarchical 3D segmentation method and the definition of vertebral body coordinate systems for QCT of the lumbar spine. Medical Image Analysis. 2006; 10(4):560–77. https://doi.org/10.1016/j.media.2006.05.005 PMID: 16828329

20. Press WH, Teukolsky SA, Vetterling WT, Flannery BP. Numerical recipes in C: Citeseer; 1996.

21. Daguet E, Jolivet E, Bousson V, Boutron C, Dahmen N, Bergot C, et al. Fat content of hip muscles: an anteroposterior gradient. The Journal of Bone & Joint Surgery. 2011; 93(20):1897–905.

22. Prodanov D, Heeroma J, Marani E. Automatic morphometry of synaptic boutons of cultured cells using granulometric analysis of digital images. Journal of neuroscience methods. 2006; 151(2):168–77. https://doi.org/10.1016/j.jneumeth.2005.07.011 PMID: 16157388

23. Sarkar N, Chaudhuri B. An efficient differential box-counting approach to compute fractal dimension of image. Systems, Man and Cybernetics, IEEE Transactions on. 1994; 24(1):115–20.

24. Glüer C-C, Blake G, Lu Y, Blunt B, Jergas M, Genant H. Accurate assessment of precision errors: how to measure the reproducibility of bone densitometry techniques. Osteoporosis International. 1995; 5 (4):262–70. PMID: 7492865

25. Ashburner J, Friston KJ. Voxel-based morphometry—the methods. Neuroimage. 2000; 11(6):805–21.

26. Ohshima S, Yamamoto S, Yamaji T, Suzuki M, Mutoh M, Iwasaki M, et al. Development of an automated 3D segmentation program for volume quantification of body fat distribution using CT. Nihon Hoshasen Gijutsu Gakkai zasshi. 2008; 64(9):1177–81. PMID: 18840956

27. Senseney J, Hemler PF, McAuliffe MJ, editors. Automated segmentation of computed tomography images. CBMS; 2009.

28. Popuri K, Cobzas D, Jagersand M, Esfandiari N, Baracos V, editors. FEM-based automatic segmentation of muscle and fat tissues from thoracic CT images. Biomedical Imaging (ISBI), 2013 IEEE 10th International Symposium on; 2013: IEEE.

29. Zhao B, Colville J, Kalaigian J, Curran S, Jiang L, Kijewski P, et al. Automated quantification of body fat distribution on volumetric computed tomography. Journal of computer assisted tomography. 2006; 30 (5):777–83. https://doi.org/10.1097/01.rct.0000228164.08968.e8 PMID: 16954927

30. Goodpaster BH, Chomentowski P, Ward BK, Rossi A, Glynn NW, Delmonico MJ, et al. Effects of physical activity on strength and skeletal muscle fat infiltration in older adults: a randomized controlled trial. Journal of Applied Physiology. 2008; 105(5):1498–503. https://doi.org/10.1152/japplphysiol.90425. 2008 PMID: 18818386

31. Kramer HR, Fontaine KR, Bathon JM, Giles JT. Muscle density in rheumatoid arthritis: Associations with disease features and functional outcomes. Arthritis & Rheumatism. 2012; 64(8):2438–50.

32. Kim JH, Choi SH, Lim S, Lim JY, Kim KW, Park KS, et al. Thigh muscle attenuation measured by computed tomography was associated with the risk of low bone density in community-dwelling elderly population. Clinical endocrinology. 2013; 78(4):512–7. https://doi.org/10.1111/cen.12016 PMID: 22901286

33. Engelke K, Adams JE, Armbrecht G, Augat P, Bogado CE, Bouxsein ML, et al. Clinical use of quantitative computed tomography and peripheral quantitative computed tomography in the management of osteoporosis in adults: the 2007 ISCD Official Positions. Journal of Clinical Densitometry. 2008; 11 (1):123–62. https://doi.org/10.1016/j.jocd.2007.12.010 PMID: 18442757

34. Arnold BA. Hybrid calibration of tissue densities in computerized tomography. US Patent 20,030,095,695; 2003.

35. GuÐjónsdóttir J, Ween B, Olsen DR. Optimal use of AEC in CT: a literature review. Radiologic technology. 2010; 81(4):309–17. PMID: 20207787

ORIGINAL RESEARCH • MUSCULOSKELETAL IMAGING

Three-dimensional Distribution of Muscle and Adipose Tissue of the Thigh at CT: Association with Acute Hip Fracture

Alexander Mühlberg, MSc • *Oleg Museyko, PhD* • *Valérie Bousson, MD* • *Pierre Pottecher, MD* •
Jean-Denis Laredo, MD • *Klaus Engelke, PhD*

From the Institute of Medical Physics (A.M., O.M., K.E.) and Department of Medicine 3 (K.E.), University Hospital, Friedrich-Alexander University Erlangen-Nuremberg, Henkestr 91, Erlangen 91052, Germany; and Department of Radiology, AP-HP, Hôpital Lariboisière and Université Paris Diderot, Paris, France (V.B., P.P., J.D.L.). Received May 11, 2018; revision requested July 2; final revision received August 27; accepted September 26. **Address correspondence to** K.E. (e-mail: *klaus.engelke@imp.uni-erlangen.de*).

The EFFECT study originally had been supported by a European Union grant (QLK6-CT-2002-02440-3DQCT). The additional analysis presented in this article was in part supported by the FORMOSA grant of the Bayerische Forschungsstiftung (1044-12).

Conflicts of interest are listed at the end of this article.

Radiology 2019; 290:426–434 • https://doi.org/10.1148/radiol.2018181112 • Content codes: CT MK

Purpose: To evaluate determinants of hip fracture by assessing soft-tissue composition of the upper thigh at CT.

Materials and Methods: In this retrospective analysis of prospectively collected data, CT studies in 55 female control participants (mean age, 73.1 years ± 9.3 [standard deviation]) were compared with those in 40 female patients (mean age, 80.2 years ± 11.0) with acute hip fractures. Eighty-seven descriptors of the soft-tissue composition were determined. A multivariable best subsets analysis was used to extract parameters best associated with hip fracture. Results were adjusted for age, height, and weight. Results of soft-tissue parameters were compared with bone mineral density (BMD) and cortical bone thickness. Areas under the receiver operating characteristic curve (AUCs) adjusted for multiple comparisons were determined to discriminate fracture.

Results: The hip fracture group was characterized by lower BMD, lower cortical thickness, lower relative adipose tissue volume of the upper thigh, and higher extramyocellular lipid (EML) surface density. The relative volume of adipose tissue combined with EML surface density (model S1) was associated with hip fracture (AUC, 0.85; 95% confidence interval [CI]: 0.78, 0.93), as well as trochanteric trabecular BMD combined with neck cortical thickness (model B2) (AUC, 0.84; 95% CI: 0.75, 0.92). The model including all four parameters provided significantly better (*P* < .01) discrimination (AUC, 0.92; 95% CI: 0.86, 0.97) than model S1 or B2.

Conclusion: In addition to bone mineral density and geometry of the proximal femur, the amount of adipose tissue of the upper thigh and the distribution of the adipocytes in the muscles are significantly associated with acute hip fracture at CT.

©RSNA, 2018

Online supplemental material is available for this article.

Prediction of osteoporotic fracture, and of hip fracture in particular, remains challenging (1). Areal bone mineral density (BMD) determined at dual x-ray absorptiometry (DXA) is the current standard of reference to determine fracture risk, but its power is insufficient to predict whether an individual will ultimately sustain a fracture. Advanced imaging methods such as quantitative CT and finite element analysis have not substantially improved hip fracture prediction (2,3).

Almost all osteoporotic hip fractures are caused by a fall; thus, it seems tempting to integrate fall-related risk factors into fracture prediction concepts. Falls are favored by muscle dysfunction or aging (4–6), but it is still unknown whether there is a direct cause-and-effect relationship between muscle function and hip fracture. Some quantitative CT studies have targeted muscle composition and size, surrogates of muscle function, as potential contributors to hip fracture risk (7–11), partly on the basis of the fact that muscle function is related to muscle fat infiltration (12–14).

In an earlier CT study, cross-sectional area of total fat, of extensor and adductor muscles, and CT attenuation of adductor muscles remained significantly lower in patients with hip fractures after adjustment for age and body mass index. In the Health, Aging, and Body Composition, or Health ABC, Study (10), decreased CT attenuation of the thigh muscle predicted hip fracture with a relative risk of 1.4 after adjustment for areal BMD. Relative risk for areal BMD alone was not reported.

In the Osteoporotic Fractures in Men Study, or MrOS, composite descriptors calculated from single-section peripheral quantitative CT measurements of the stress strain index, the bending strength of bone and muscle cross-sectional area, were determined at 66% of the tibial length. The descriptors predicted hip fracture with a hazard rate of up to 1.2 after adjustment for areal BMD of the spine or total hip (7). The hazard rate for areal BMD alone was 2.1.

In addition to muscle density and respective volumes of macroscopic adipose tissue and muscle in different anatomic compartments, other factors may contribute to

Abbreviations

AUC = area under the ROC curve, BMD = bone mineral density, CI = confidence interval, DXA = dual x-ray absorptiometry, EFFECT = European Femur Fracture Study, EML = extramyocellular lipid, IMAT = intermuscular adipose tissue, OR = odds ratio, PAT = perimuscular adipose tissue, ROC = receiver operating characteristic, SAT = subcutaneous adipose tissue, VOI = volume of interest

Summary

The amount and distribution of muscle and adipose tissue of the upper thigh is related to the risk of hip fracture independent of bone mineral density of the hip.

Implications for Patient Care

- Hip fracture prediction may be improved by investigation of soft-tissue descriptors of the thigh in combination with standard bone mineral density assessments.
- Advanced muscle texture descriptors are independently associated with hip fracture.

hip fracture risk. For example, lipid aggregation patterns change with higher age (15). In congenital muscle dystrophies, radiologists semiquantitatively grade the muscle appearance using morphologic criteria such as "washed out" and "moth eaten," appearances characterized, respectively, by low density and high porosity (16–19). In muscle degenerative diseases, the Goutallier scoring system is most commonly used to semiquantitatively grade the relative amount of fatty and muscle tissue densities (20,21).

In our present study, we used a number of structure descriptors to quantify the soft-tissue distribution. We hypothesized that descriptors quantifying a moth-eaten and washed-out appearance are associated with hip fracture risk. We applied this combined analysis to the cross-sectional European Femur Fracture Study (EFFECT), in which quantitative CT studies of elderly women were obtained immediately after hip fracture and in a control group (22,23) using the same statistical procedures as before. Further details are given in Appendix E1 (online).

Materials and Methods

For the prospective cross-sectional EFFECT investigation, ethics approval was obtained from the local ethics committees from the two recruiting centers in Paris, France, and Manchester, England. All patients gave written informed consent.

Study Population

The study population and CT acquisition process have been described previously (22). The subpopulation of EFFECT used in the present study consisted of 40 Caucasian postmenopausal women with acute hip fractures (mean age, 80.2 years ± 11.0 [standard deviation]) and 55 control subjects (mean age, 73.1 years ± 9.3) who had volun-

teered to participate in the study. Exclusion criteria were previous hip fracture, but not a history of other fractures or treatment with osteoporotic medications. In Paris, 28 patients with fractures and 24 control participants were recruited, and in Manchester, 12 patients with fractures and 31 control participants were recruited. Twenty-one patients had a fracture of the left leg, and 19 had a fracture of the right leg. In the patients with fractures, the nonfractured leg was analyzed, and in the control participants, the left leg was analyzed. Patients with fractures were scanned before surgery, usually within 24 hours after the fracture. Further details, including a flowchart, are presented in Appendix E1 (online).

Tissue Composition of the Thigh

The fascia separates subcutaneous adipose tissue (SAT) from the intrafascial volume of interest (VOI) (Fig 1). The intrafascia contains anatomic muscles and perimuscular adipose tissue (PAT). Within anatomic muscles, adipose tissue is either contained in adipocytes, embedded among muscle fibers (extramyocellular lipids [EMLs]), or stored as droplets in the myocytes (intramyocellular lipids [IMLs]). Adipocytes are typically arranged in clusters. Depending on cluster size and the spatial resolution of the CT image, EMLs can be directly segmented because their CT values differ from that of the surrounding muscle tissue. IMLs can be measured only indirectly through CT muscle tissue density, which is lower with higher IML values (24). Muscle tissue is defined as muscle minus EMLs, while IMLs are part of muscle tissue (Fig 1). In our present study, we refer to intermuscular adipose tissue (IMAT) as a combination of PAT and EML (Fig 1), which does not include IML (25). This is in agreement with the traditional conception of IMAT as the "fat signal intensity, both within and between the muscles" visible on non–fat suppressed T1-weighted MR images (26).

Image Analysis

A global VOI of the upper femoral thigh including both legs that extended 3 cm distally from the maximum extension of the lesser trochanter was analyzed. This VOI excluded the dermis, genital organs, and bone. Details were described

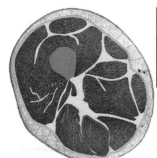

Muscle-Lipid System (MLS)			
Intrafascia (IF)			Subcutaneous Adipose Tissue (SAT)
Anatomical Muscle (M)		PAT	
Muscle Tissue (MT) including IML	EML		
	IMAT		

Figure 1: Muscle, adipose tissue, and lipids in the thigh. The color scheme used in the thigh corresponds to the tissue labels in the table at right. The femur is shown in green. *EML* = extramyocellular (intramuscular) lipids, *IMAT* = intermuscular adipose tissue (a combination of EMLs and perimuscular adipose tissue [PAT]), *IML* = intramyocellular lipids (part of muscle tissue), *MT* = muscle tissue (ie, anatomic muscle without EMLs but including IMLs).

previously (27). In short, the fascia was first segmented by using a model-based approach to separate SAT from the intrafascia volume. Anatomic muscle was obtained as a complement of PAT. Muscle tissue was segmented with the help of a Gaussian mixture model (28) applied to the histogram of all voxels of the global VOI. For anatomic muscle and muscle tissue, mean CT values and standard deviations were obtained. The percentage of adipose tissue volume within the global VOI of the upper thigh was also determined by using a Gaussian mixture model (27).

Author (A.M.), who had developed the method, performed all analyses. Intra- and interoperator reanalysis precision errors of volumes of the analysis VOIs denoted in Figure 1 were less than or equal to 0.5% (27). Therefore, a second independent analysis was not performed.

A second segmentation strategy used the gray-value histogram of all voxels within the intrafascia to specify binary VOIs of six different muscle and lipid/adipose tissue compositions (Fig 2) (29). Details are described in Appendix E2 (online). In brief, a lower intensity threshold T_{AT} determined from the gray-value distribution of the SAT VOI defined pure adipose tissue (ie, a muscle concentration of 0%). An upper threshold T_M, determined in a group of 30 young healthy male athletes (age, 30.0 years ± 3.1), defined a muscle concentration of 100%

(high-density muscle). Muscle concentrations between these two extremes were equally divided into four "bins" (Fig 2).

After segmentation, standard measurements such as volume and density, plus a number of structure descriptors describing the morphology of the muscle-lipid system, were determined (Appendix E3 [online]). Inhomogeneity, undirectedness, variogram slope, and gray-level structure tensor were determined for nonbinary VOIs. Euler characteristics, surface density, and grain size were determined for binary VOIs. For clarity, the descriptors are denoted here by VOI descriptor name (eg, EML-Local Undirectedness). Densities were determined as the average CT value of a given VOI measured in Hounsfield units after appropriate calibration (Appendix E2 [online]).

Statistical Analysis

Independent variables adjusted for age, height, and weight were excluded from multivariable analysis if they did not discriminate between patients with fractures and patients without fractures in univariable analysis (with analysis of covariance). Independent variables were further excluded if their values differed significantly (paired t test) between the right and left legs of the control participants. If independent variables correlated highly ($r > 0.8$) with each other, those with lower statistical significance (larger P values) also were not further processed.

Multivariable best subsets selection for binary logistic regression with adjustments for age, height, and weight was performed for the set of remaining descriptors to identify independent correlates of the presence of hip fractures. Five best subsets models were used for the detailed binary logistic regression analysis. Odds ratios (ORs) were normalized for a one-standard-deviation change of the test variable and are reported with 95% confidence intervals (CIs). The Hosmer-Lemeshow test was used to evaluate model robustness.

For comparison, only absolute ORs are reported, even if the change was negative. The area under the curve (AUC) obtained from a receiver operating characteristic (ROC) analysis was selected as a discriminative performance characteristic of the binary logistic regression models and was reported with 95% CIs. The DeLong test was used to compare AUCs among models (30). The outcome of the DeLong test is conservative, in that it is valid if the difference is significant but not if the difference is not significant (31). If the DeLong test resulted in nonsignificance, the improvement of a given model was evaluated by the significance of additional parameters adjusted for the original set of parameters (31,32). The statistical analysis was performed by using IBM SPSS Statistics 23 and R software (version 3.3.2, www.R-project.org). $P \leq .05$ was considered to indicate a significant difference.

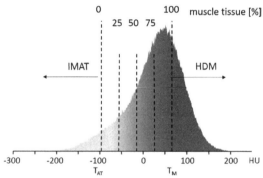

Figure 2: Histogram used to binarize the intrafascia volume of interest (VOI) into high-density muscle, intermuscular adipose tissue *(IMAT)*, and four additional VOIs containing voxels with 0%–25% muscle tissue, 75%–100% adipose tissue, et cetera. HDM = high-density muscle (defined as a muscle concentration of 100%), T_{AT} = threshold for adipose tissue, T_M = threshold for HDM. The shading codes the CT value in Hounsfield units from −100 HU for IMAT to purple for HDM. CT values in between represent different contributions of muscle and adipose tissue in a given voxel for which a color between yellow and purple is assigned.

Table 1: Participant Demographic Data

Parameter	Group with Fracture (n = 40)	Group without Fracture (n = 55)	P Value
Mean age ± standard deviation (y)	80 ± 11 (58–100)	73 ± 9 (60–95)	.002
Mean height ± standard deviation (cm)	157 ± 7 (143–175)	157 ± 7 (138–171)	>.9
Mean weight ± standard deviation (kg)	59 ± 12 (33–85)	65 ± 12 (34–98)	.04

Note.—Data in parentheses are ranges. Between control and fracture groups, only age was significantly different ($P < .001$).

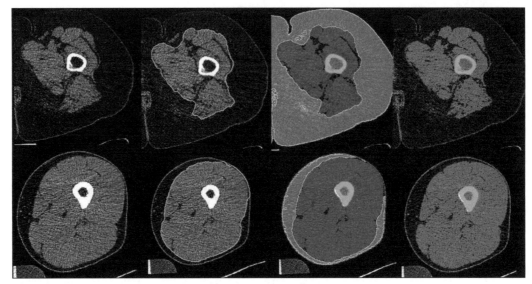

Figure 3: Representative axial CT sections. Top: Images in 62-year-old female participant in the European Femur Fracture Study, or EFFECT. Bottom: Images in a younger, healthy 44-year-old woman for comparison. From left to right: Unpro cessed CT images; segmented fascia shown as yellow contours; subcutaneous adipose tissue (yellow), muscle (purple), femoral bone (green), perimuscular adipose tissue (uncolored); and muscle tissue (purple).

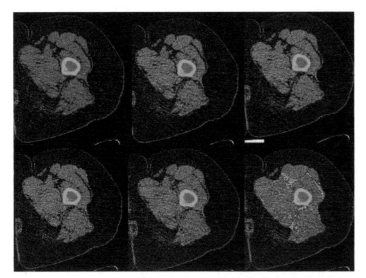

Figure 4: Axial CT sections show binary volumes of interest (VOIs). Voxels belonging to a binary VOI are shown in color using the color scheme of Figure 2. Top left: High-density muscle (100% muscle tissue); top center: 80% muscle tissue and 20% lipids; top right: 60% muscle tissue and 40% lipids; bottom left: 40% muscle tissue and 60% lipids; bottom center: 20% muscle tissue and 80% lipids; bottom right: intermuscular adipose tissue. Each VOI shows the distribution of CT density values categorized into six different bins based on the histogram shown in Figure 2. Green = femoral bone.

Results

Study Population

Patient demographics are shown in Table 1. Compared with the earlier analysis (22), in seven participants with fractures and in five control participants, the CT field of view was too small for the thigh soft-tissue analysis. As reported before, the hip fracture group pooled from both centers was older (P = .002) than the pooled control group.

Segmented Images

Figure 3 shows CT images of the thigh in an elderly (62-year-old) participant in EFFECT and a younger healthy woman (age 44 years) for comparison. Figure 4 shows the six binary VOIs inside the fascia, starting with high-density muscle (top left) to IMAT (bottom right). Figure 5 shows a variety of muscle lipid distributions in three EFFECT participants.

Univariable Analysis

Of 87 soft-tissue descriptors, 24 were associated with hip fracture in comparison to the control participants (P ≤ .05 for all). After adjustment for age, height, and weight, 16 significant discriminators remained. Four pairs of this subset were highly correlated (r > 0.8). The final set of 12 descriptors (Table 2) was included in the multivariable analysis.

Nine of the 12 descriptors characterized texture. They are described in detail in Appendix E3 (online). The muscle descriptors, muscle peak and muscle standard deviation, were derived from the Gaussian mixture model applied to all voxels inside the intrafascia. They denote the CT value (in Hounsfield units) of the peak and standard deviation of the muscle distribution. A low CT value of the peak describes a distribution that is shifted toward smaller CT values, indicating a higher lipid infiltration within muscle tissue. The CT value of the muscle peak differs from muscle density, which is the mean CT value of all voxels of the anatomic muscle.

Multivariable Analysis

Table 3 shows the five best subsets models, models S1–S5, for fracture discrimination. They are ordered according to the Bayesian information criterion (33) of the best subsets procedure, which combines the number of variables and goodness-of-fit of the binary regression model. When age was added as an independent descriptor, the descriptor combinations of the five models remained the same. Therefore, age, height, and weight were not considered as independent descriptors but were used as covariates.

The AUC was very similar for all five models, which was not surprising, as global percentage of adipose tissue contributed to all models and EML-Surface Density contributed to four out of the five models. The third descriptor in models S2, S4, and S5 was not significant. Replacing global percentage of adipose tissue with the relative volume of SAT resulted in lower AUCs.

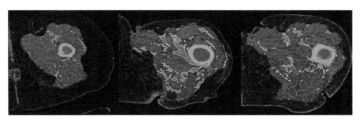

Figure 5: Axial CT sections show a variety of muscle lipid distributions showing the binary volumes of interest (VOIs) of high-density muscle (purple) and intermuscular adipose tissue (yellow) in three different participants in the European Femur Fracture Study, or EFFECT. The four other binary VOIs were omitted. Green = femoral bone.

For example, adjusted AUCs of model S1 were reduced from 0.85 to 0.8, as shown in Table 3.

For comparison, Table 4 shows results for global percentage of adipose tissue alone, as this was the dominant descriptor in all five multivariable models. Table 4 also shows results for two quantitative CT models suggested earlier (23) that were recalculated for the population in our present study. Model B1 contained integral BMD of the total hip only, which is very easy to measure. Model B2 contained a combination of trabecular BMD of the trochanter and cortical thickness of the neck, one of the strongest quantitative CT models for hip fracture prediction (23).

Table 4 also shows results for three combinations of bone and soft-tissue models. AUCs for the combination of soft-tissue descriptors (model S1) with the quantitative CT models B1 or B2 were larger than those for model B1 (P = .01 for adjusted models) or model B2 (P = .02, adjusted) alone. The combination of model S1 with trochanteric trabecular BMD and neck cortical thickness (model B2) resulted in higher AUCs than those for model S1 alone (P = .03, adjusted). A combination of model B1 or B2 with S2–S5 models instead of model S1 resulted in very similar results.

The combination of global percentage of adipose tissue alone with model B2 resulted in AUCs that were still higher than those for either quantitative CT model (significantly for model B1, with P = .007 adjusted) but were lower than those for the combination of models S1 and B2 where the

Table 2: Results of Univariable Analysis of Muscle and Adipose Tissue Descriptors

Volume of Interest and Muscle and Adipose Tissue Descriptor	Group without Fracture (n = 55)*	Group with Fracture (n = 40)*	Mean Difference[†]	Unadjusted P Value[‡]
Global percentage of adipose tissue	59.4 ± 4.7	54.3 ± 6.3	−5.1 (−8.6)	<.001 (<.001)
Intrafascial				
Muscle peak CT value (HU)	42.2 ± 7.7	37.3 ± 5.7	−4.9 (−11.6)	.001 (.03)
Standard deviation of distribution of muscle CT values (HU)	51.1 ± 8.6	52.6 ± 11.0	1.0 (1.7)	.46 (.008)
Muscle				
Variogram slope (gv)	2.2 ± 0.3	1.9 ± 0.3	−0.2 (−10.2)	<.001 (.05)
Gray-value structure tensor angle Z (radians)	1.5 ± 0.1	1.5 ± 0.1	0.02 (0.9)	.41 (.01)
Muscle tissue				
Euler density (betti/cm³)	9.6 ± 5.8	11.4 ± 7.7	1.8 (18.6)	.25 (.23)
Mean grain size (mm³)	5.0 ± 1.4	4.3 ± 1.0	−0.69 (−13.7)	.005 (.04)
High-density muscle				
Local undirectedness (gv)	79.6 ± 1.2	79.9 ± 1.1	0.33 (0.41)	.36 (.01)
Euler density (betti/cm³)	−86 ± 49	−113 ± 41	−29 (−34)	.006 (.04)
Extramyocellular lipids				
Local undirectedness (gv)	78.9 ± 0.9	79.3 ± 1.1	0.43 (0.5)	.05 (.002)
Local inhomogeneity (gv)	0.019 ± 0.07	0.023 ± 0.013	0.004 (19)	.15 (.008)
Surface density (mm)	5.0 ± 0.5	5.3 ± 0.6	0.3 (5.9)	.005 (.02)

Note.—Analysis of covariance was performed of the 12 muscle and adipose tissue descriptors that differed significantly between patients with fracture and control participants. gv = Gray value.

* Data are means ± standard deviations.

[†] Data in parentheses are percentage mean differences.

[‡] Data in parentheses are P values adjusted for age, height, and weight.

Table 3: Best Subsets Models of Soft-Tissue Parameters

Model No., Volume of Interest, and Descriptor	Unadjusted OR*	Adjusted OR*	P Value†	Model AUC‡	Hosmer-Lemeshow P Value§
S1				0.81 (0.73, 0.90)/ 0.85 (0.78, 0.93)	.8 (.61)
Global percentage of adipose tissue	3.9 (2.0, 7.6)	4.5 (2.1, 10.0)	<.001 (<.001)		
Extramyocellular lipid surface density	2.8 (1.5, 5.2)	2.6 (1.3, 5.2)	.001 (.006)		
S2				0.81 (0.73, 0.90)/ 0.85 (0.78, 0.93)	.34 (.36)
Global percentage of adipose tissue	3.5 (1.8, 6.8)	4.3 (1.9, 9.5)	<.001 (<.001)		
Intrafascial muscle peak attenuation	1.3 (0.7, 2.4)	1.2 (0.6, 2.3)	.4 (.7)		
Extramyocellular lipid surface density	2.3 (1.2, 4.6)	2.4 (1.1, 5.1)	.01 (.02)		
S3				0.80 (0.71, 0.89)/ 0.84 (0.76, 0.92)	.059 (.16)
Global percentage of adipose tissue	3.4 (1.8, 6.3)	4.1 (1.9, 8.8)	<.001 (<.001)		
Muscle tissue mean grain size	2.2 (1.3, 3.8)	2.1 (1.2, 4.0)	.003 (.02)		
S4				0.82 (0.73, 0.90)/ 0.85 (0.77, 0.93)	.85 (.11)
Global percentage of adipose tissue	3.7 (1.9, 7.2)	4.5 (2.0, 10.0)	<.001 (<.001)		
Muscle variogram slope	1.3 (0.7, 2.4)	1.2 (0.6, 2.7)	.4 (.6)		
Extramyocellular lipid surface density	2.4 (1.2, 4.7)	2.4 (1.1, 5.1)	.01 (.03)		
S5				0.81 (0.73, 0.90)/ 0.86 (0.78, 0.93)	.68 (.42)
Global percentage of adipose tissue	3.9 (2.0, 7.6)	4.5 (2.1, 10.0)	<.001 (<.001)		
Muscle tissue mean grain size	1.4 (0.7, 2.7)	1.4 (0.6, 3.0)	.4 (.4)		
Extramyocellular lipid surface density	2.3 (1.0, 5.0)	2.1 (0.9, 4.9)	.05 (.09)		

Note.—Best subsets models were obtained from descriptors in Table 2. Results were adjusted for age, height, and weight. P values are shown with and without adjustment for age, height and weight. AUC = area under the receiver operating characteristic curve, OR = odds ratio.

* Data in parentheses are 95% confidence intervals (CIs).

† Data in parentheses are P values adjusted for age, height, and weight.

‡ Data are unadjusted AUCs with 95% CIs in parentheses/adjusted AUCs with 95% CIs in parentheses. AUCs were adjusted for age, height, and weight.

§ For the Hosmer-Lemeshow test, $P > .05$ indicates that the null hypothesis (observed event rates match expected event rates in subgroups of the model population) is not rejected.

muscle component was added (P = .21, adjusted). Figure 6 shows the corresponding ROC curves.

Discussion

Our study combined descriptors of BMD, bone geometry, muscle, adipose tissue, and intramuscular lipid distribution to determine association with hip fracture risk. Compared with the best bone model consisting of trochanteric trabecular BMD and cortical thickness of the neck, association with hip fracture was significantly improved (P = .02) when combined with lower relative volume of adipose tissue within the global VOI and higher surface density of the adipocyte distribution within muscle (measured as EML surface density). Corresponding AUCs adjusted for age, weight, and height were 0.84 for the bone model and 0.92 for the combined model. Surprisingly, association with hip fracture for soft-tissue models alone (AUC, 0.84–0.86) was as strong as for the best bone model alone (AUC, 0.84). The percentage volume of adipose tissue of the upper thigh was the dominant soft-tissue contributor (AUC,

0.81), but descriptors of the muscle-lipid distribution contributed significantly to the soft-tissue models ($P < .02$) and the combined soft-tissue/bone models ($P < .01$).

The descriptors explored in the present study can be classified into three types of contributors to fracture or fracture protection: First, bone strength, which is highly correlated to BMD and cortical thickness, second, mechanical protection against hip fracture by a larger cushion reflected by the relative volume of adipose tissue of the global VOI, and third, muscle degeneration expressed by the surface density of the adipocyte distribution within muscle. The cause-and-effect relationships of most soft-tissue descriptors with fracture reduction still have to be evaluated, but the importance of adipose tissue for risk prediction of hip fracture has been shown before in the Health ABC Study (10). In a Chinese female population, the global percentage of adipose tissue showed an adjusted AUC of 0.83, similar to our results of 0.81 (9). Hip protectors are also based on the idea of adding a cushion around the hip, similar to a higher amount of SAT (34).

Interestingly, in our present study, muscle density, volume of SAT, and IMAT, which were often measured in other

Table 4: Best Subsets Models Combining Bone and Soft-Tissue Parameters

Type of Model, Volume of Interest, and Descriptor	Unadjusted OR*	Adjusted OR*	P Value[†]	Model AUC[‡]
G: Global percentage of adipose tissue as single descriptor	2.9 (1.7, 5.0)	3.6 (1.8, 7.4)	<.001 (<.001)	0.74 [0.63, 0.85]/ 0.81 (0.71, 0.90)
Models containing quantitative CT descriptors—BMD and cortical geometry selected from the literature (23)				
B1				0.76 (0.66, 0.86)/ 0.79 (0.70, 0.88)
Total hip integral BMD	3.1 (1.7, 5.5)	2.9 (1.4, 6.0)	<.001 (.004)	
B2				0.79 (0.70, 0.88)/ 0.84 (0.75, 0.92)
Trochanter trabecular BMD	2.5 (1.4, 4.6)	3.9 (1.7, 8.5)	.002 (.001)	
Neck cortical thickness	1.9 (1.1, 3.4)	1.7 (0.8, 3.4)	.04 (.2)	
Combination of "bone" and "muscle and adipose tissue" models				
G and B2				
Global percentage of adipose tissue	2.8 (1.5, 5.2)	4.1 (1.8, 9.2)	.001 (.001)	0.84 (0.76, 0.92)/ 0.89 (0.83, 0.96)
Trochanter trabecular BMD	2.6 (1.4, 5.1)	4.2 (1.7, 10.2)	.003 (.001)	
Neck cortical thickness	1.8 (1.0, 3.3)	1.8 (0.8, 4.0)	.07 (.1)	
S1 and B1				0.87 (0.79, 0.94)/ 0.89 (0.82, 0.96)
Global percentage of adipose tissue	3.5 (1.8, 6.93)	4.44 (2.0, 10.0)	<.001 (<.001)	
Extramyocellular lipid surface density	2.4 (1.3, 4.61)	2.3 (1.1, 4.8)	.009 (.02)	
Total hip integral BMD	2.5 (1.3, 4.84)	2.9 (1.2, 6.6)	.005 (.02)	
S1 and B2				0.89 (0.83, 0.96)/ 0.92 (0.86, 0.97)
Global percentage of adipose tissue	3.6 (1.8, 7.29)	4.5 (1.9, 10.7)	<.001 (.001)	
Extramyocellular lipid surface density	2.7 (1.4, 5.32)	2.3 (1.1, 5.0)	.004 (.04)	
Trochanter trabecular BMD	2.3 (1.2, 4.37)	4.0 (1.6, 10.0)	.02 (.004)	
Neck cortical thickness	2.1 (1.1, 3.99)	1.8 (0.8, 4.1)	.03 (.1)	

Note.—Table shows associations with hip fracture, comparing the power of percentage of adipose tissue alone with that of "bone" quantitative CT models B1 and B2 and combinations of model S1 (described in Table 3) with models B1 and B2. AUC = area under the receiver operating characteristic curve, BMD = bone mineral density, OR = odds ratio.

* Data in parentheses are 95% confidence intervals (CIs).

[†] Data in parentheses are P values adjusted for age, height, and weight.

[‡] Data are unadjusted AUCs with 95% CIs in parentheses/adjusted AUCs with 95% CIs in parentheses. AUCs were adjusted for age, height, and weight.

CT muscle studies (7,8,10,24), did not enter the multivariable analysis. Mean CT values (ie, densities of anatomic muscle and muscle tissue) were significantly different between the fracture and control groups in our study but not after adjustment for age, height, and weight. Relative volumes of IMAT, muscle, and muscle tissue did not significantly differ at all. Average muscle volume (2.6%) and muscle tissue volume (1%) were just slightly higher in the control group. A similar finding concerning muscle density was reported in the study with the Chinese female population mentioned above (9). CT values between control subjects and participants with fractures differed only for the adductor muscles and not for the other muscle groups. In the prospective Health ABC Study, muscle density and area of the thigh predicted fracture risk after adjustment for age, height, body mass index, and some additional covariates, but only muscle density remained after an additional adjustment for DXA BMD with a hazard rate of 1.5 (10). One important difference from other studies may be the three-dimensional

analysis applied in our study. Cross-sectional muscle and adipose tissue area obtained from the analysis of single sections may show larger variations between groups than volume.

The innovative aspect of our study was the use of parameters characterizing the muscle-lipid distribution, which showed that muscle in patients with fracture was more "moth eaten" and "washed out." Details are outlined in Appendix E3 (online). Limitations of our study were the cross-sectional design, the small number of patients, and the age difference between the two groups. No adjustments were performed for prior fractures or use of antiosteoporosis medication. Because of the moderate age difference of 7 years between the two groups, all results were age adjusted. Age matching would have reduced the number of patients to less than 50. Also, the interpretation of the structure descriptors, while plausible, remains speculative, and further investigations linking structure descriptors to the muscle-lipid distribution are required. Finally, to reduce the large number of descriptors that univariately were associated with fracture, a somewhat subjective exclusion procedure was used. However, this procedure reduced the number of input descriptors for the best subsets analysis to 12, a reasonable number given that fewer than 100 patient data sets were available. Finally, the DeLong test used to compare performance between models underestimates significance, as already stated in the Statistical Analysis section. Thus, our statistical approach was very conservative in comparison to current methods used in medical image analysis (35).

In conclusion, our study results show that association with hip fracture can be improved if relative adipose tissue volume and descriptors of the adipocyte distribution within muscle are combined with BMD and geometry. BMD and relative adipose tissue volume were the strongest contributors, while cortical thickness and EML-Surface Density still contributed significantly, although with lower ORs. Soft-tissue descriptors provided a significant independent contribution to fracture discrimination.

Finally, if the results of this study are confirmed in larger populations, it will be highly relevant to add soft-tissue descriptors, in particular the measurement of percentage adipose tissue volume, to dual x-ray absorptiometry (DXA), the current reference standard for bone mineral density. Whether local DXA measurements can be used for a combined assessment remains an interesting question.

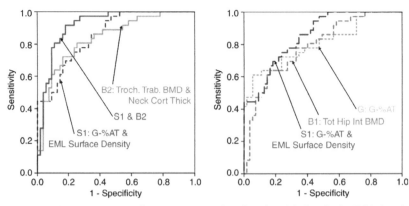

Figure 6: Receiver operating characteristic curves for selected models described in Table 3 and Table 4. Left: Comparison of model S1 (combination of percentage adipose tissue in the global volume of interest [VOI]—that is, G-%AT—and surface density of the extramyocellular lipid [EML] VOI) and model B2 (combination of trochanter trabecular bone mineral density [Troch. Trab. BMD] and cortical thickness of the neck [Neck Cort Thick]) with the combined S1 and B2 model. Right: Comparison of models G (percentage adipose tissue in the global VOI) and B1 (total hip integral BMD [Tot Hip Int BMD]) with model S1.

Author contributions: Guarantors of integrity of entire study, A.M., K.E.; study concepts/study design or data acquisition or data analysis/interpretation, all authors; manuscript drafting or manuscript revision for important intellectual content, all authors; manuscript final version approval, all authors; agrees to ensure any questions related to the work are appropriately resolved, all authors; literature research, A.M., J.D.L., K.E.; clinical studies, V.B., P.P., J.D.L.; experimental studies, A.M., J.D.L.; statistical analysis, O.M., K.E.; and manuscript editing, A.M., V.B., J.D.L., K.E.

Disclosures of Conflicts of Interest: A.M. disclosed no relevant relationships. **O.M.** disclosed no relevant relationships. **V.B.** disclosed no relevant relationships. **P.P.** disclosed no relevant relationships. **J.D.L.** Activities related to the present article: disclosed no relevant relationships. Activities not related to the present article: paid by Sanofi for advice on steroid injection products; paid by Pfizer, BMS, Abbvie, and Amgen for lectures on the radiographic signs of various diseases; has received travel and meeting funds from Guerbet to attend RSNA 2017 and 2018. Other relationships: disclosed no relevant relationships. **K.E.** Activities related to the present article: disclosed no relevant relationships. Activities not related to the present article: is a part-time employee of Bioclinica; institution has received a grant from the European Approach Project. Other relationships: disclosed no relevant relationships.

References

1. Rubin KH, Friis-Holmberg T, Hermann AP, Abrahamsen B, Brixen K. Risk assessment tools to identify women with increased risk of osteoporotic fracture: complexity or simplicity? a systematic review. J Bone Miner Res 2013;28(8):1701–1717.
2. Engelke K, Lang T, Khosla S, et al. Clinical use of quantitative computed tomography (QCT) of the hip in the management of osteoporosis in adults: the 2015 ISCD official positions—part I. J Clin Densitom 2015;18(3):338–358.
3. Zysset P, Qin L, Lang T, et al. Clinical use of quantitative computed tomography-based finite element analysis of the hip and spine in the management of osteoporosis in adults: the 2015 ISCD official positions—part II. J Clin Densitom 2015;18(3):359–392.
4. Reis P, Moro A, Bins Ely V, et al. Universal design and accessibility: an approach of the influence of muscle strength loss in the risk of falls in the elderly. Work 2012; 41(Suppl 1):374–379.
5. Wang X, Ma Y, Wang J, et al. Mobility and muscle strength together are more strongly correlated with falls in suburb-dwelling older Chinese. Sci Rep 2016;6(1):25420.
6. Wickham C, Cooper C, Margetts BM, Barker DJ. Muscle strength, activity, housing and the risk of falls in elderly people. Age Ageing 1989;18(1):47–51.
7. Wong AK, Cawthon PM, Peters KW, et al. Bone-muscle indices as risk factors for fractures in men: the Osteoporotic Fractures in Men (MrOS) Study. J Musculoskelet Neuronal Interact 2014;14(3):246–254.
8. Wong AK, Beattie KA, Min KK, et al. Peripheral quantitative computed tomography-derived muscle density and peripheral magnetic resonance imaging-derived muscle adiposity: precision and associations with fragility fractures in women. J Musculoskelet Neuronal Interact 2014;14(4):401–410.
9. Lang T, Koyama A, Li C, et al. Pelvic body composition measurements by quantitative computed tomography: association with recent hip fracture. Bone 2008;42(4):798–805.
10. Lang T, Cauley JA, Tylavsky F, et al. Computed tomographic measurements of thigh muscle cross-sectional area and attenuation coefficient predict hip fracture: the health, aging, and body composition study. J Bone Miner Res 2010;25(3):513–519.
11. Hahn MH, Won YY. Bone mineral density and fatty degeneration of thigh muscles measured by computed tomography in hip fracture patients. J Bone Metab 2016;23(4):215–221.
12. Frank-Wilson AW, Chalhoub D, Figueiredo P, et al. Associations of quadriceps torque properties with muscle size, attenuation, and intramuscular adipose tissue in older adults. J Gerontol A Biol Sci Med Sci 2018;73(7):931–938.
13. Frank-Wilson AW, Farthing JP, Chilibeck PD, et al. Lower leg muscle density is independently associated with fall status in community-dwelling older adults. Osteoporos Int 2016;27(7):2231–2240.
14. Inacio M, Ryan AS, Bair WN, Prettyman M, Beamer BA, Rogers MW. Gluteal muscle composition differentiates fallers from non-fallers in community dwelling older adults. BMC Geriatr 2014;14(1):37.
15. Kirkland JL, Tchkonia T, Pirtskhalava T, Han J, Karagiannides I. Adipogenesis and aging: does aging make fat go MAD? Exp Gerontol 2002;37(6):757–767.
16. Kornblum C, Lutterbey G, Bogdanow M, et al. Distinct neuromuscular phenotypes in myotonic dystrophy types 1 and 2 : a whole body highfield MRI study. J Neurol 2006;253(6):753–761.
17. Mercuri E, Talim B, Moghadaszadeh B, et al. Clinical and imaging findings in six cases of congenital muscular dystrophy with rigid spine syndrome linked to chromosome 1p (RSMD1). Neuromuscul Disord 2002;12(7-8):631–638.
18. Swash M, Brown MM, Thakkar C. CT muscle imaging and the clinical assessment of neuromuscular disease. Muscle Nerve 1995;18(7):708–714.
19. Wattjes MP, Kley RA, Fischer D. Neuromuscular imaging in inherited muscle diseases. Eur Radiol 2010;20(10):2447–2460.
20. Goutallier D, Postel JM, Bernageau J, Lavau L, Voisin MC. Fatty muscle degeneration in cuff ruptures: pre- and postoperative evaluation by CT scan. Clin Orthop Relat Res 1994;(304):78–83.
21. Oh JH, Kim SH, Choi JA, Kim Y, Oh CH. Reliability of the grading system for fatty degeneration of rotator cuff muscles. Clin Orthop Relat Res 2010;468(6):1558–1564.
22. Bousson VD, Adams J, Engelke K, et al. In vivo discrimination of hip fracture with quantitative computed tomography: results from the prospective European Femur Fracture Study (EFFECT). J Bone Miner Res 2011;26(4):881–893.
23. Museyko O, Bousson V, Adams J, Laredo JD, Engelke K. QCT of the proximal femur: which parameters should be measured to discriminate hip fracture? Osteoporos Int 2016;27(3):1137–1147.

24. Goodpaster BH, Kelley DE, Thaete FL, He J, Ross R. Skeletal muscle attenuation determined by computed tomography is associated with skeletal muscle lipid content. J Appl Physiol (1985) 2000;89(1):104–110.

25. Addison O, Marcus RL, Lastayo PC, Ryan AS. Intermuscular fat: a review of the consequences and causes. Int J Endocrinol 2014;2014:309570.

26. Karampinos DC, Baum T, Nardo L, et al. Characterization of the regional distribution of skeletal muscle adipose tissue in type 2 diabetes using chemical shift-based water/fat separation. J Magn Reson Imaging 2012;35(4):899–907.

27. Mühlberg A, Museyko O, Laredo JD, Engelke K. A reproducible semi-automatic method to quantify the muscle-lipid distribution in clinical 3D CT images of the thigh. PLoS One 2017;12(4):e0175174.

28. Reynolds D. Gaussian mixture models. In: Li S, Jain A, eds. Encyclopedia of biometrics. New York, NY: Springer, 2009; 659–663.

29. Daguet E, Jolivet E, Bousson V, et al. Fat content of hip muscles: an anteroposterior gradient. J Bone Joint Surg Am 2011;93(20):1897–1905.

30. DeLong ER, DeLong DM, Clarke-Pearson DL. Comparing the areas under two or more correlated receiver operating characteristic curves: a nonparametric approach. Biometrics 1988;44(3):837–845.

31. Demler OV, Pencina MJ, D'Agostino RB Sr. Misuse of DeLong test to compare AUCs for nested models. Stat Med 2012;31(23):2577–2587.

32. Vickers AJ, Cronin AM, Begg CB. One statistical test is sufficient for assessing new predictive markers. BMC Med Res Methodol 2011;11(1):13.

33. Schwarz GE. Estimating the dimension of a model. Ann Stat 1978;6(2):461–464.

34. Parkkari J, Kannus P, Heikkilä J, Poutala J, Sievänen H, Vuori I. Energy-shunting external hip protector attenuates the peak femoral impact force below the theoretical fracture threshold: an in vitro biomechanical study under falling conditions of the elderly. J Bone Miner Res 1995;10(10):1437–1442.

35. Erickson BJ, Korfiatis P, Akkus Z, Kline TL. Machine learning for medical imaging. RadioGraphics 2017;37(2):505–515.

SEMPAI: a Self-Enhancing Multi-Photon Artificial Intelligence for Prior-Informed Assessment of Muscle Function and Pathology

Alexander Mühlberg,* Paul Ritter, Simon Langer, Chloë Goossens, Stefanie Nübler, Dominik Schneidereit, Oliver Taubmann, Felix Denzinger, Dominik Nörenberg, Michael Haug, Sebastian Schürmann, Roarke Horstmeyer, Andreas K. Maier, Wolfgang H. Goldmann, Oliver Friedrich, and Lucas Kreiss

Deep learning (DL) shows notable success in biomedical studies. However, most DL algorithms work as black boxes, exclude biomedical experts, and need extensive data. This is especially problematic for fundamental research in the laboratory, where often only small and sparse data are available and the objective is knowledge discovery rather than automation. Furthermore, basic research is usually hypothesis-driven and extensive prior knowledge (priors) exists. To address this, the Self-Enhancing Multi-Photon Artificial Intelligence (SEMPAI) that is designed for multiphoton microscopy (MPM)-based laboratory research is presented. It utilizes meta-learning to optimize prior (and hypothesis) integration, data representation, and neural network architecture simultaneously. By this, the method allows hypothesis testing with DL and provides interpretable feedback about the origin of biological information in 3D images. SEMPAI performs multi-task learning of several related tasks to enable prediction for small datasets. SEMPAI is applied on an extensive MPM database of single muscle fibers from a decade of experiments, resulting in the largest joint analysis of pathologies and function for single muscle fibers to date. It outperforms state-of-the-art biomarkers in six of seven prediction tasks, including those with scarce data. SEMPAI's DL models with integrated priors are superior to those without priors and to prior-only approaches.

1. Introduction

Artificial intelligence (AI) and especially deep learning (DL) is experiencing great success in the classification of digital image data. These algorithms are nowadays regularly used for research in the medical field for automated diagnostics in macroscopic imaging, such as computed tomography (CT) or magnetic resonance imaging (MRI), for example to estimate the effectiveness of radiation therapy,[1] to automatically phenotype COPD,[2] or to segment organs.[3]

For fundamental research in the laboratory (lab) and in animal models, however, the automation aspect is much less relevant. Instead, lab research is more concerned with basic discoveries that can lead to a better understanding of pathology or function. Specifically for animal models and fundamental research with microscopy, the mere automation of disease detection does not add much value, since trained DL algorithms require translation to humans. In addition, lab experiments are often of small

A. Mühlberg, P. Ritter, S. Nübler, D. Schneidereit, M. Haug,
S. Schürmann, O. Friedrich, L. Kreiss
Institute of Medical Biotechnology
Department of Chemical and Biological Engineering
Friedrich-Alexander University Erlangen-Nuremberg
Paul-Gordan-Str. 3, 91052 Erlangen, Germany
E-mail: alexander.mueale.muehlberg@fau.de

P. Ritter, D. Schneidereit, O. Friedrich, L. Kreiss
Erlangen Graduate School in Advanced Optical Technologies
Paul-Gordan-Str. 6, 91052 Erlangen, Germany

S. Langer, O. Taubmann, F. Denzinger, A. K. Maier
Pattern Recognition Lab
Department of Computer Science
Friedrich-Alexander University Erlangen-Nuremberg
Martensstr. 3, 91058 Erlangen, Germany

C. Goossens
Clinical Division and Laboratory of Intensive Care Medicine
KU Leuven
UZ Herestraat 49 – P.O. box 7003, Leuven 3000, Belgium

D. Nörenberg
Department of Radiology and Nuclear Medicine
University Medical Center Mannheim
Medical Faculty Mannheim
Theodor-Kutzer-Ufer 1–3, 68167 Mannheim, Germany

The ORCID identification number(s) for the author(s) of this article can be found under https://doi.org/10.1002/advs.202206319

DOI: 10.1002/advs.202206319

sample size and labels are sparse, thus rather data-hungry DL should not be employed to mitigate overfitting, and the lab researcher therefore has to rely on hypothesis-based research, often in combination with statistics. However, it would be helpful to use hypotheses, while having a system that can recognize patterns independently, e.g., via convolutional neural networks (CNN) that identify relevant features automatically. The field of optical microscopy, in particular, has already benefited from a broad variety of AI applications,[4] such as automation,[5,6] segmentation,[7] and image quality (IQ) enhancement including optimal illumination,[8] emitter localization in super-resolution microscopy,[9] or image restoration.[10] However, current research mostly covers AI optimization of the microscope settings and is less focused on the hypothesis-based approach of small-scale experiments for biological knowledge discovery. A scientist conducting basic research in the lab is also often unfamiliar with the selection of an appropriate DL architecture and associated hyperparameter tuning, which can also be a limiting factor for the use of AI in the lab.

All of these points justify the need for an AI that is designed specifically to meet the needs of a biomedical lab researcher. To understand how this can be achieved, we briefly introduce two cutting-edge areas of research: meta-learning and the integration of prior knowledge.

Meta-learning, or "learning to learn", analyzes that conditions must be given to be able to effectively learn a specific task. This includes the relatively new field of neural architecture search (NAS),[11] with the goal to automatically identify a suitable NN architecture for a given problem. Meta-learning might replace the time-consuming trial-and-error process of manual architecture search and may not only provide competitive performance, but also solutions with particularly desirable properties, such as curiosity.[12] On the downside, NAS, and more generally meta-learning, are computationally expensive approaches, although a variety of techniques are developed to decrease time and associated costs.[13] Recently, a novel meta-learning approach for segmentation problems in biomedical imaging gained a lot of attention: nnU-Net³. nnU-Net optimizes NN architecture and hyperparameters together with rule-based image processing operations (normalization, resampling etc.), with the eponymous U-Net serving as the base NN architecture. This approach outperformed most prevailing methods for many automated segmentation problems in biology and medicine.[3]

And although DL has shown its strengths for big data, e.g., for automated classification of images in the world wide web, for fundamental medical research with limited data sets, methods based on prior knowledge can show competitive performance for describing or predicting a pathology.[14] Providing prior biological knowledge, or in brief "priors", to the learning algorithm as a baseline instead of starting from scratch, therefore, seems plausible. Another common drawback of many DL systems is the lack of explainability. A large number of methods, such as DeepSHAP,[15] are developed to highlight the image information relevant for the decision-making process. However, a fundamental question posed by Rudin[16] was why the current research focuses on post-hoc explanations of complicated models rather than creating more interpretable models from the beginning. Explainability can be increased by using priors, such as established measurements or known biomarkers, in the learning process of a NN. Additionally, the integration of prior knowledge in the form of known operators as NN layers was already shown to stabilize the learning process by reducing the maximum error bounds.[17] Lastly, by integration of biological priors, human understanding and intuition about a problem can be employed within AI research. Modern AI approaches for microscopy also already started to integrate prior knowledge of imaging physics. For instance, the integration of physics knowledge into the learning process of an AI helped with the technological optimization of microscope- and software-components[18,19] for enhanced IQ, and with digital staining of virtual fluorescence in label-free phase microscopy,[20] or Fourier ptychography microscopy.[21]

Based on the considerations regarding an AI for the lab, and the cutting-edge areas of meta-learning and prior-integration discussed above, we present the Self-Enhancing Multi-Photon Artificial Intelligence (SEMPAI) that is specifically designed to integrate hypothesis-driven priors in a meta-learning approach for fundamental research. SEMPAI as a general tool simultaneously identifies optimal data representation, degree of prior integration, and NN architecture for a given biomedical problem. In contrast to the technologically-inspired optimization of microscope parameters for enhanced IQ, it performs biologically-inspired meta-learning, i.e., the optimization in a biologically interpretable configuration space, on already existing databases for knowledge discovery. Additionally, SEMPAI utilizes multi-task learning over different tasks to leverage common patterns shared over all prediction tasks to also enable the prediction for small and sparse data sets. Lastly, SEMPAI's models that are trained on a large joint database with different pathologies in animal models could then be used as foundation models,[22,23] and be fine-tuned for, e.g., translation from ex vivo to in vivo experiments or from animal models to humans. Summarizing, SEMPAI aims to integrate the hypotheses of researchers and identify biologically relevant information in experiments of low sample size, simultaneously serving as a generator of foundation models based on a large database.

To demonstrate the value of this approach, we apply SEMPAI to an extensive and unique multi-study data collection of 1,298 3D second-harmonic generation (SHG) images from isolated muscle fibers of different morphologic, genetic, pathologic, or functional conditions. Images of the database were acquired with label-free multiphoton microscopy (MPM), and functional parameters with highly automated robotized biomechatronics systems.

R. Horstmeyer, L. Kreiss
Computational Optics Lab
Department of Biomedical Engineering
Duke University
101 Science Dr, Durham, NC 27708, USA

W. H. Goldmann
Biophysics Group
Department of Physics
Friedrich-Alexander University Erlangen-Nuremberg
Henkestr. 91, 91052 Erlangen, Germany

2. Results

2.1. SEMPAI Method Overview

In the context of this publication, priors are handcrafted features, i.e., either already known imaging biomarkers or novel features that were developed based on the researcher's hypotheses. Labels, as usual, define the values to be learned and predicted.

SEMPAI simultaneously optimizes configurations of its three main components: the prior integration, the data representation (DaRe), and NN architecture and hyperparameters (NN settings). This self-enhancement in a biologically interpretable configuration space is logged, and its evaluation enables knowledge discovery. The method is shown in **Figure 1**, the configuration space in **Table 1**, an extended rationale and explanations for the configuration space, as well as details about the implementation, in Methods.

SEMPAI can choose from five different levels to integrate priors (or hypotheses). It can learn without priors (*NoPriors*), use them as auxiliary tasks (*AuxLosses*), which results in a soft constraint to the learning problem[1], integrate them as an additional branch into the fully connected layer of the NN (*Branches*), or a combination of both (*AuxLosses&Branches*). In the fifth configuration (*PriorsOnly*), only priors are used in an integrated AutoML method[24] for handcrafted features, i.e., without using the SHG images and DL. To the best of our knowledge, the integration within SEMPAI is the first attempt to combine current priors (biomarkers) known in single fiber muscle research with ML. Further extended explanations are provided in Methods.

The decisions by SEMPAI regarding DaRe indicate "how and where" biological information can optimally be learned. For example, SEMPAI analyzes whether *3D* images are needed or whether three regularly spaced representative slices (*2.5D*) are sufficient and how large this spacing should be. Analogously,

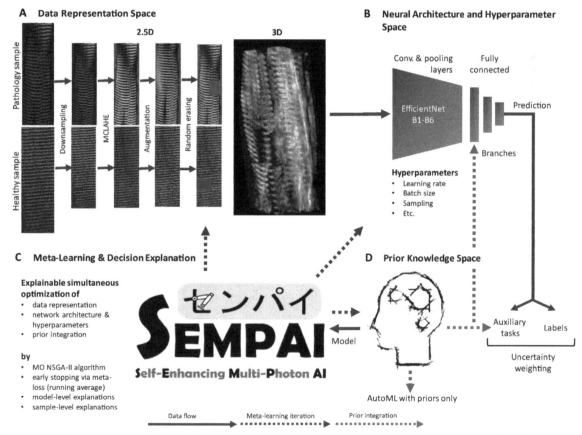

Figure 1. SEMPAI method overview. For each iteration, termed trial, of-the-self-enhancement process, a data representation (DaRe) is selected that represents the images either in 2.5D, i.e., by three regularly spaced slices, or in 3D. Then, decisions are made regarding the modification of the DaRe such as downsampling or contrast enhancement. The selected DaRes are fed to a NN, and the NN architecture and its hyper-parameters are selected. The level of prior integration is then chosen. SEMPAI decides, whether priors i) are not employed, ii) are used as auxiliary tasks for the NN training, iii) are fed directly to the fully connected layer of the NN as branches, or iv) are used in both integration methods, i.e., a combination of (ii) and (iii). In option v), the priors are the only input to an AutoML approach for handcrafted features. The resulting model of the trial is used to predict the labels on the dev set. The performance of the model yields the meta-losses that guide SEMPAI's configuration selection for the next trial. This process results in a simultaneous self-enhancement of DaRe, NN architecture & hyperparameters and prior integration with increasing number of trials. Table 1 shows the configuration space. Scale bar: 25 μm.

**ADVANCED
SCIENCE NEWS**
www.advancedsciencenews.com

**ADVANCED
SCIENCE**
Open Access
www.advancedscience.com

Table 1. Configuration space. Decisions made by SEMPAI during self-enhancement process.

Data Representation Variants				
Contrast Enhancement	*Yes*: the MCLAHE algorithm is applied on the images	*No*: No further enhancement after registration and resampling		
Down-sampling	*Yes*: Images are downsampled to 0.75µm voxel size isotropically	*No*: No resampling, 0.5 µm isotropically		
Augmentation	*Yes*: Application of 3D augmentation such as Gaussian noise, rotation, flipping, affine transformation	*No*: Original standardized images are used		
Random Erasing	*Yes*: Random regions of the image are erased	*No*: Original standardized images are used		
Volume/slice selection	*3D*: The whole 3D array of each sample is used	*2.5D_1*	*2.5D_5*	*2.5D_10*	*2.5D_20*
		Center slice and 2 slices with 1, 5, 10 or 20 µm distance to the center slice are selected		

Prior Integration Variants				
Method	*NoPriors*	*AuxLosses*	*Branches*	*AuxLosses&Branches*	*PriorsOnly*

NN Settings Variants	
Capacity	2D/3D EfficientNet *B1-B6*
Learning rate	Cyclic (*Yes/No*) and in range [0.0001, 0.2]
Optimizer	Adam or SGD with Nesterov moment
Momentum	Momentum in range [0.9, 0.99]
Gradient Clipping	*Yes*: gradients are clipped to the norm 1.0 — *No*: NN gradients evolve freely
Batch Size	2.5D/3D: *small* (32/4); *medium* (64/8), *large* (128/16), *XL* [256/32], *XXL* (512/64)
Imbalance Sampling	*Yes*: class distributions are re-balanced based on strata information of the initial train-dev-test split. Sampling weights are estimated automatically — *No*: The original data distribution is fed in the NN

SEMPAI provides information on the importance of downsampling, which can help to estimate the required image resolution for a learning task. As a side effect, this feedback may also have an impact on future studies. For instance, if SEMPAI finds that larger pixel sizes are sufficient for a given task, future data could be acquired in shorter scan times, increasing experimental throughput.

For NN training, in addition to hyperparameter optimization, SEMPAI selects one architecture variant from the base architecture EfficientNet[25] that offers scaled variants with different capacity (*B1* to *B6*) and for 2D or 3D, and has been shown to yield competitively predictive performance with less DOF than alternative architectures.[25] Accordingly, this architecture allows relatively fast training, making it advantageous for utilization with time-consuming meta-learning. SEMPAI learns all tasks jointly in a multi-task setup. Our hypothesis is that this enforces a semantic regularization of the learning process, since systematic differences unrelated to the biological origin, e.g., in IQ, are less likely to be used for prediction. Instead, the use of related muscle-specific patterns across different learning tasks is enforced. Multi-task learning further has the advantage that tasks with small data can still be learned, as DOF are determined by information from similar tasks with more data.[26] Recent research shows that this joint learning is preferable to the similar concept of transfer learning.[27] In case of missing labels for either primary or auxiliary losses of a sample, no backpropagation occurs during training for the corresponding model outputs, i.e., these outputs are "masked" for that sample. This results in a sort of interleaved learning, in which different tasks are learned in different batches. It also enables joint training without the need for data imputation, thereby enabling effective analysis of sparse and heterogeneous laboratory data. During NN training, all losses are weighed against each other by uncertainty weighting.[28]

Data to be analyzed by SEMPAI are split into training (train), development (dev) and test set (more details in 2.3). The train set is used to train the NN, while the dev set is used to optimize SEMPAI's decisions in the configuration space. The test set remains unseen. The resulting model of each trial created on the train set is applied for prediction of the labels on the dev set. The predictive performance of the chosen model for each task is assessed for the dev set. Those performances are used as meta-losses to select the configurations for the next trial. SEMPAI uses NSGA-II[29] multi-objective optimization for this selection, i.e., there is not only one loss to be minimized, but the losses of all labels are minimized independently.

For tasks with small data, i.e., pCa50 and passive force (**Table 2** in the next Section 2.2), ML and especially DL are severely limited due to overfitting. Based on the identified associations of the same priors with the investigated labels in previous studies,[30–34] we hypothesize that muscle-specific learning tasks are related and the mean predictive performance over all tasks may assist SEMPAI to select an even more regularized model. Therefore, a *total meta-loss* is introduced, which is a weighted sum of all meta-losses for each task and provides an estimate of the model performance over all tasks. This loss is not used for optimization, but for selection of models for small data tasks (N<100).

**ADVANCED
SCIENCE NEWS**

www.advancedsciencenews.com

**ADVANCED
SCIENCE**
Open Access

www.advancedscience.com

Table 2. Used multi-study data after standardization and exclusion of data with inappropriate IQ. Please note that the total number of unique images is not a sum of all above, since most images had information for multiple labels, e.g., an image from mdx, where active force was available. Example images for each of the included studies are shown in Supporting Information 1. WT: wild type, C: classification, R: regression.

Label / Task	Data set with reference	Total number of curated images
Inflammatory phenotype: Sepsis/WT, C	A[30]	731
Dystrophic phenotype: *mdx*/WT, C	B1[31], B2[31], C[31], D[32]	567
Muscle type: Diaphragm/EDL, C	D[32]	179
Active Force, R	B1[31]	232
Active Force/pCa, R	B1[31]	152
Passive (Restoration) Force, R	C[31]	39
pCa50, R	B2[31]	39
Total number of unique images		**1,298**

SEMPAI explains itself regarding i) decision-making during the self-enhancement process (SEMPAI model-level explanations) as well as regarding ii) the decision-relevant image pixels/voxels and priors of each sample (SEMPAI sample-level explanations): i) For each task, based on the performed experiments and their results, SEMPAI retrospectively fits a random forest model to estimate its predictive performance from a given configuration. Subsequently, the fitted model is fed in the SHAP Tree Explainer[35] to estimate the impact of DaRe, NN settings, and prior integration and identify configurations that yield models with good predictive performance. ii) For the sample-level explanation of important image regions, SEMPAI utilizes Deep SHAP.[15] In the case of prior integration method *Branches* or combined *AuxLosses&Branches*, the method was extended to provide attribution of priors together with, and orthogonally to, the attribution map of the image.

2.2. Database, Priors, and Labels

To apply SEMPAI, a database for an organ has to be fed in. Without compromising the general approach of the method, in this section we present and describe the specific database analyzed by SEMPAI. The standardization of this database is explained in the next Section 2.3.

We retrospectively screened in-house experimental MPM data of single muscle fibers acquired over more than a decade to obtain a large database that includes a variety of biological properties with respect to muscle pathology and function. For these data, we utilize current MPM imaging biomarkers as priors while pathological and functional parameters serve as labels, which we describe below.

A variety of muscle pathologies affect a structured muscle morphology, leading to reduced function of the entire system. For instance, Duchenne muscular dystrophy (DMD) results in an overall loss of structural integrity in individual fibers, eventually leading to failure of respiratory and heart muscle that can be life-limiting.[36] Besides chronic degenerative diseases like DMD, also acute myopathies can result in disruptions of the myofib-

rillar structural alignment, as it has been shown in ongoing sepsis.[30]

Function of muscle tissue is based on its passive mechanical and its active force generation properties. Passive force parameters are related to the visco-elastic behavior of the muscle. In contrast, active force parameters describe its intrinsic ability to generate force, e.g., represented by the physiological sensitivity to calcium ions. Automated integrated biomechatronics systems, such as the *MyoRobot*[37,38] or the *MechaMorph* system,[31] can measure these active and passive parameters simultaneously to imaging of the fiber. Both aforementioned systems consist of force transducers (FT) to measure force and voice coil actuators (VCA) to perform axial movement with higher precision as compared to stepper motors.[39] A combination of high-resolution label-free SHG microscopy with biomechanical measurements of active and passive force was recently demonstrated.[31] Through this, correlations between morphological features derived from SHG and functional properties acquired with FTs and VCAs were experimentally shown for individual muscle fibers from *mdx* and wild type (WT) mice.

Table 2 shows the investigated learning tasks, i.e., the labels, the corresponding original studies, and the number of samples used. The extended variants, i.e., larger sample size, of the following studies are included in our database: A) For investigating muscle atrophy during sepsis, samples from the *extensor digitorum longus* (EDL) of septic and WT mice were imaged and complemented by active force recordings in EDL single fibers of the same animal.[30] Sepsis is used here as a surrogate for the inflammatory phenotype. B1) Active force measurement and subsequent SHG imaging at each force recording were carried out in EDL single fibers from WT and *mdx* mice.[31] B2) Force recordings from a different image data set of EDL fibers to deduce the Ca^{2+} sensitivity of the contractile apparatus, pCa50, as a measure for the troponin-C Ca^{2+} sensor characteristics.[38,40] C) The same setting was used to access passive force parameters on a different set of animals.[31] D) Fixated single fibers and fiber bundles from EDL and diaphragm in *mdx* and WT animals were imaged to investigate structural differences between *mdx* and WT as well as between the muscle types of EDL and diaphragm.[32] Here, the *mdx* mice serves as a surrogate for the dystrophic phenotype.

The 3D images presented in this data set were generated by label-free SHG microscopy. Compared to other, macroscopic label-free imaging modalities, such as MRI,[41] CT,[42] or ultrasound,[43] SHG imaging provides sub-μm resolution to resolve sarcomeres ($\approx$2 μm in size), which is relevant to establish a deeper understanding of the structure-function relationship and the impact of pathologies on single muscle fibers.[44,45] From these images, morphological image features were computed with previously reported software.[46] In brief, these features include the *cosine angle sum* (CAS) taken from selected 2D planes (2D-CAS) and in 3D (3D-CAS), the *vernier density* (VD), the 3D sarcomere length (3D-SL), and the cross-sectional area (CSA) of single fibers. Since these features have already been shown to be descriptive for a variety of rather specific remodeling patterns in muscle research, related to aging, chronic degenerative or inflammatory myopathies,[33–34] we use them as priors.

A more elaborate explanation of the image acquisition, the robotized biomechatronics system, and the extraction of priors is provided in Methods.

2.3. Cross-Study Data Standardization, Data Split, and Performance Metrics

The workflow for data acquisition, standardization, and the resulting data distribution after splitting in train, dev, and test set is shown in **Figure 2**.

The original experiments were conducted by different experimenters, with different imaging systems and parameters, resulting in a high degree of data heterogeneity. Thus, standardization is required to compare images from different studies acquired under varying experimental conditions and during different time periods. By standardization, the technical variance can be minimized. In brief, the images are resampled to an isotropic voxel size of 0.5 μm, slightly denoised via a median filter, and the background, which is defined by Otsu's thresholding of the image, is set to zero. Then, the Multidimensional Contrast Limited Adaptive Histogram Equalization (MCLAHE) algorithm[47] is applied for contrast enhancement of each muscle fiber. This contrast-enhanced image is registered to a pre-selected fiber with canonical orientation and fiber pattern by a rigid multi-scale registration, and the resulting transformation is applied to the non-enhanced version. A mean image of the registered fibers is created after setting all foreground voxels to one, which provides the probability of presence for muscle fibers in each voxel. The bounding box of voxels with probability >0.85 is generated (extent: $180 \times 80 \times 57\ \mu m^3$) and applied to the images. This cropping of images to relevant regions enables the use of DL models with reduced degrees of freedom (DOF), which is advantageous for our data regime. To allow a uniform comparison of the CSA between each of the individual experiments, we developed a method that combines three different variants of CSA estimation to detect outliers and to be more robust. Further fine-grained details about the implementation of the standardization and the CSA estimation are provided in Methods.

For our meta-learning, the standardized data are stratified and grouped into train, dev, and test set (2/4, 1/4, 1/4). The grouping prevents different images of single fibers extracted from the same muscle bundle from being distributed over different sets, which would result in information leakage. The stratification ensures that the distribution in the respective sets is similar, thus, label instances with rare occurrence are present in the train, dev, and test sets. The label and prior data are normalized to a standard score based on mean and standard deviation of the train set. The predictive performance is given for the unseen test set (holdout) as area under the curve (AUC) of the receiver operating characteristic for classification tasks and as R^2 for regression tasks.

2.4. SEMPAI as a Tool for Fundamental Knowledge Discovery

As described above, to explain its decision-making on the model-level, SEMPAI computed the respective SHAP values of the samples and the mean absolute SHAP values over all samples to quantify the association of the configuration space with the predictive performance. In addition, the stability of the analysis was tested (see Methods). The results are shown in **Figure 3**A.

In five of seven investigated tasks, the level of prior integration was the most important decision. For the classification tasks *mdx*, sepsis, and muscle type, the integration of prior knowledge

(or hypotheses) was especially important according to the mean absolute SHAP values. While *mdx* and muscle type preferred the soft constraint of priors as *AuxLosses*, the harder learning task of predicting sepsis preferred a stronger integration of priors as *AuxLosses&Branches*. Although selecting the level of prior integration was on average not the most important decision for learning muscle function, the highest positive impact on predictive performance was found with strong prior integration, namely *Branches* and *AuxLosses&Branches*, for active and passive force. The results for pCa50 are harder to interpret. According to the individual SHAP values, the task preferred no prior integration or weak integration as *AuxLosses* but the pattern of the association is rather complex. Further explanation, for instance, why a prior integration *AuxLoss* can be considered weaker than Branches, can be found in Methods.

Most learning tasks, especially *mdx* and sepsis, benefited from smaller NN capacity, indicating that fewer DOF were sufficient for the complexity of the task and helped to avoid overfitting.

Reducing image resolution had a negative impact on five of seven learning tasks, although at varying degrees, as indicated by the SHAP values when employing down-sampling. Especially *mdx* profited from a higher resolution. Those tasks required detailed information of highly resolved muscle filament structures, while especially sepsis worked better at downsampled resolution, indicating that future imaging data could be recorded at higher throughput for this task.

Prediction of active and passive force benefited from contrast enhancement. This is also intuitively comprehensible when inspecting the images visually, as the IQ for function assessment is lower on average due to a more complex experimental setup[31] (Supporting Information S1). On the contrary, the modification of image intensities by contrast enhancement had a negative effect for the tasks *mdx*, muscle type and sepsis. This indicates that not only the structure, but also the original intensity yields important information for these tasks and should not be artificially modified.

In two of seven tasks, the selection of the spacing between three representative slices was the most important decision. Interestingly, for active and passive force prediction, SEMPAI strongly profited from using slices from the periphery of the muscle fiber (±20 μm), compared to using further slices in the proximity of the muscle center (i.e., 1, 5, and 10 μm). This indicates additional biological information for function assessment in the muscle periphery in comparison to a sole evaluation of the muscle center. Using configurations that employed 3D DaRe generally provided an inferior predictive performance, and none of these models was found among the best 100 for any task.

mdx was the only task for which localized properties were of special interest, since the performance decreased by employing random erasing.[48] We explain this technique, and how we use it, in Methods. In brief, random erasing is an augmentation method that regularizes by preventing a model from using only one or a small number of image regions to learn. Random erasing of a few image regions prevents this, similar to the concept of dropout[49] for neurons. However, if an information occurs only locally, random erasing leads to a situation where prediction is no longer possible. To investigate this effect for *mdx*, we used the sample-level decision explanation of SEMPAI. This confirmed that, in addition to 2D-VD, localized regions of twisted or damaged muscle

**ADVANCED
SCIENCE NEWS**

www.advancedsciencenews.com

**ADVANCED
SCIENCE**
Open Access

www.advancedscience.com

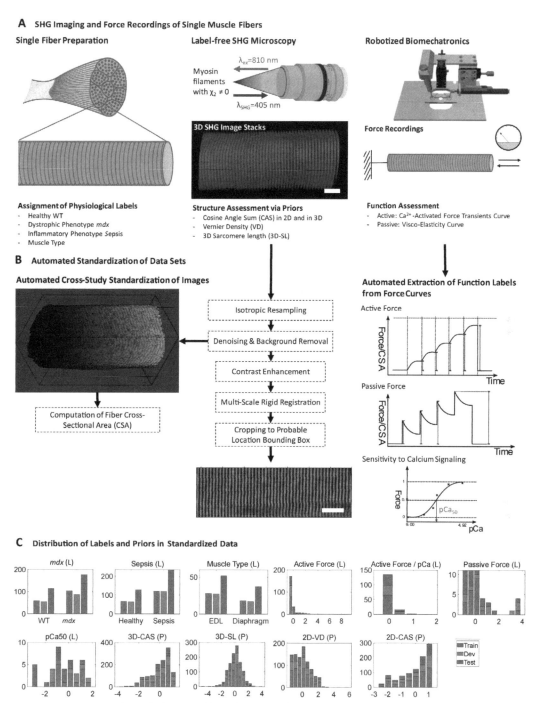

Figure 2. Data acquisition, cross-study standardization, and value distribution of labels and priors in train, dev, and test set. A) Single muscle fibers were dissected from murine muscle tissue. The data were annotated regarding pathologies and muscle type. In each case, 3D label-free second harmonic generation (SHG) microscopy was performed, and morphological features, termed priors, were calculated. Muscle tissue was assessed for its function by robot-assisted biomechanical force measurements. B) The SHG images are standardized with a dedicated image processing pipeline consisting of resampling, denoising, registration, contrast enhancement and cropping of the images to a probable location bounding box. Within this standardization process, the cross-sectional area (CSA) of the fibers is calculated. Function labels like active force or pCa50 are automatically computed from the raw curves coming from the biomechatronics system. C) Distribution of priors (P) and labels (L) in train, development (dev) and test data after stratified grouped data split. The distributions are normalized to standard score. Scale bar: 25 μm.

**ADVANCED
SCIENCE NEWS**
www.advancedsciencenews.com

**ADVANCED
SCIENCE**
Open Access
www.advancedscience.com

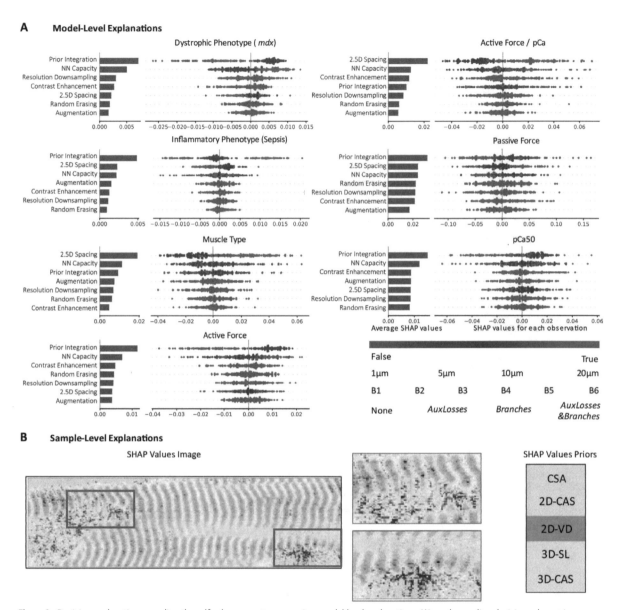

Figure 3. Decision explanation regarding the self-enhancement process, i.e., model-level explanations (A), and regarding decision-relevant image voxels/pixels and priors, i.e., sample level-explanations (B). A) A random forest model learns the predictive performance of SEMPAI for a specific label as a function of the configuration space. The resulting model is then analyzed by SHAP Tree Explainer that allows to estimate the individual contribution of each configuration for each sample in units of the performance metrics (AUC/R^2). Decisions are sorted top-to-bottom based on their mean absolute SHAP values as a surrogate for the importance of the decision. Configurations are color-coded from weak to strong expression of a configuration (legend in lower right). B) Attribution map of image (left) and priors (right) for one mdx sample. Colored voxels and priors are used by SEMPAI for this sample to correctly predict *mdx*. The attribution of priors is computed simultaneously and shown with the same color code and scale.

fibers were especially used to predict *mdx*. An example of such a finding is shown in (Figure 3B). When those image regions were randomly erased, a loss of predictive performance was observed. For all other tasks, however, more global properties seem to be important for prediction, making the augmentation effect[48] of random erasing more advantageous.

As demonstrated for *mdx*, SEMPAI provides a detailed sample-level highlighting of important image regions, orthogonally to the information given by priors. A collection of examples is shown in Supporting Information S2. However, for a proper quantitative evaluation, those observations must be validated in a standardized manner, which is beyond the scope of this study.

**ADVANCED
SCIENCE NEWS**
www.advancedsciencenews.com

**ADVANCED
SCIENCE**
Open Access
www.advancedscience.com

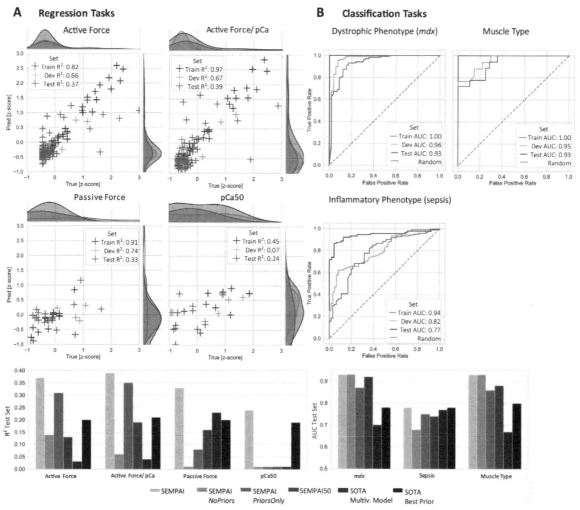

Figure 4. Overall results of SEMPAI, its sub-configurations and comparison with state-of-the-art (SOTA) methods. Performance metrics in train (for NN training), dev (for meta-optimization) and test set (unseen data) for regression A), R^2, and classification, AUC, tasks B).

In the future, an observer study based on SEMPAI could lead to novel scientific insights.

In most of its decisions, SEMPAI autonomously chose a stronger regularization. This was achieved by a strong prior integration, a low NN capacity, and also the lower dimensional 2.5D DaRe.

2.5. Predictive Performance of SEMPAI's Foundation Models and Comparison to Benchmarks

To benchmark the performance of SEMPAI, we implemented two state-of-the-art (SOTA) baselines: as univariate analyses still reflect the standard approach in laboratory research, especially in a low sample size setting, we select the best prior on the combined train and dev set and use it as a univariate predictor for the test set. In addition, to assess the performance of SOTA multivariate modeling, we use all priors and fit a statistical pipeline,

consisting of MRMR[50] feature selection, best subset selection. and multiple linear/logistic regression, on the combined train and dev set. The resulting model is applied for the prediction on the test set. For fair benchmarking, as statistical models are more severely regularized, potentially resulting in underfitting, we vary the best subset selection information criterion (Akaike/Bayesian) and the penalty of the regression (L2/elastic net: L1&L2) and report the best performance on the test set. To understand the merit of priors and images individually, we report the results of SEMPAI when using only priors (SEMPAI *PriorsOnly*), i.e., when it does not have access to the images, and the opposite, i.e., exclude trials that integrated priors (SEMPAI *NoPriors*). Finally, to test susceptibility of SEMPAI for non-optimal configurations, we give the average performance of the 50 best models (SEMPAI50).

The detailed results of SEMPAI, including train and dev set performance, and the comparison with SOTA are shown in **Figure 4** and **Table 3**. In six of seven investigated learning tasks,

ADVANCED
SCIENCE NEWS _____
www.advancedsciencenews.com

ADVANCED
SCIENCE
Open Access
www.advancedscience.com

Table 3. Overall SEMPAI results in train, dev, and test set; of SEMPAI sub-configurations, and comparison with SOTA methods (all results on the test set if not denoted otherwise). ns: negative sign, i.e., worse than guessing.

Task	SEMPAI Train/Dev/Test	SEMPAI NoPriors/ PriorsOnly	SEMPAI50	SOTA Multiv. Model	SOTA Best Prior
Mdx [AUC]	1.0/0.96/0.93	0.93/0.87	0.92	0.70	2D-VD: 0.78
Sepsis [AUC]	0.94/0.82/0.77	0.68/0.75	0.74	0.77	3D-SL: 0.77
Muscle Type [AUC]	1.0/0.95/0.93	0.93/0.86	0.88	0.67	2D-VD: 0.80
Active Force [R^2]	0.82/0.66/0.37	0.14/0.31	0.13	0.03	2D-CAS: 0.20
Active Force/pCa [R^2]	0.97/0.67/0.39	0.06/0.35	0.19	0.04	3D-CAS: 0.21
Passive Force [R^2]	0.91/0.74/0.33	ns/0.08	0.16	0.23	2D-CAS: 0.20
pCa50 [R^2]	0.45/0.07/0.24	ns/ns	ns	0.01	3D-CAS: 0.19

SEMPAI's foundation models were superior to SOTA models in predicting the labels of the test set.

Active force was predicted by SEMPAI with R^2 0.37, while SOTA gave 0.20 using the prior 2D-CAS. The prediction of the biologically more interesting active force adjusted for pCa yielded similar results with a performance of R^2 0.39 by SEMPAI and 0.21 for SOTA by prior 3D-CAS. For passive force, SEMPAI again achieved solid results with R^2 0.33, while SOTA achieved 0.23 via the multivariate model. For pCa50, SEMPAI was only slightly superior, R^2 0.24, to using the prior 3D-CAS, R^2 0.19. Predictions of force parameters were more susceptible to performance decrease for non-optimal configurations than those of pathologies and muscle type, evident from the results for SEMPAI50, which in the case of the force predictions showed inferior results compared to the best trial.

As expected, the prediction for tasks with very small sample size, pCa50 and passive force, was problematic for models with large DOF or without strict regularization as shown by the predictive performance of DL (SEMPAI *NoPriors*), single-task AutoML (SEMPAI *PriorsOnly*) and, in case of pCa50, even a simple multivariate statistical model with only few DOF. SEMPAI's regularization by multi-task learning, integration of priors, and the model selection based on the *total meta-loss*, however, resulted in a SEMPAI model with slightly improved performance compared to the best SOTA approach, the univariate predictor 3D-SL (one DOF).

In three of seven tasks, SEMPAI *PriorsOnly* was superior to SEMPAI *NoPriors* and especially achieved competitive performance in classification tasks and for predicting active force. The priors already provided the diagnostic information for classifying the inflammatory phenotype sepsis, since no improvement in predictive performance was observed by additional utilization of DL on images.

In contrast, for the dystrophic phenotype *mdx* and the muscle type, SEMPAI *NoPriors* yielded very strong models and, in the case of *mdx*, these predictions were superior to those based solely on priors. Thus, the performance of *PriorsOnly* or *NoPriors* models varied largely between tasks. In all tasks, however, SEMPAI identified a level of prior integration on the dev set that led to a good generalizability, i.e., the best predictive performance for the test set.

Especially for the prediction of muscle function, synergistic effects of combining prior knowledge with DL are observed, as

SEMPAI provided strongly improved performance compared to DL without priors or models solely based on priors. These effects may be interpreted as a DL-based prior (or hypothesis) refinement.

3. Discussion

We developed a novel Self-Enhancing Multi-Photon Artificial Intelligence (SEMPAI) and applied it on a total of 1,298 single muscle fiber 3D SHG images. SEMPAI targets close interaction with biomedical researchers. On the one hand, SEMPAI integrates, tests, and refines prior knowledge or hypotheses of the domain expert. On the other hand, SEMPAI gives systematic feedback about influencing factors for optimal extraction of biologically relevant information. The researchers can therefore use their domain knowledge as input to the method and receive comprehensible and easy-to-interpret feedback as output.

The foundation models generated by SEMPAI were superior to previous state-of-the-art (SOTA) biomarkers in predicting active and passive muscle force, pCa50 for Ca^{2+}-activated isometric force, muscular dystrophy phenotype in the *mdx* mouse as well as murine muscle type. To the best of our knowledge, deep learning (DL) was not yet applied to MPM image databases in single muscle fiber research. For muscle research, DL was for example applied to gene data from DMD patients[51] or to perform functional evaluation of DMD on ultrasound images.[52] Most often, DL in this context is used on clinical MRI data, e.g., for the identification of MRI biomarkers in smaller cohorts (N = 26),[53] for image classification[54] or for the analysis of larger clinical cohorts (N = 432).[55] However, ultrasound and MRI do not offer sufficient resolution to understand DMD and the *mdx* model at the level of individual muscle fibers. Here, MPM has the unique advantages of label-free image contrast and sub-cellular resolution. In support of the microscopic approach, SEMPAI also showed that the prediction of most of the prediction tasks benefit from finer resolution.

Usually, imaging-based biomarker studies are either purely based on priors, especially if the sample size is low like in many clinical imaging studies, or novel DL architectures. However, meta-learning on our multi-study data indicates that a prior integration, by varying degrees, in DL methods almost always yielded the best predictive performance, especially for the prediction of muscle function. Recent research, such as

**ADVANCED
SCIENCE NEWS**

www.advancedsciencenews.com

**ADVANCED
SCIENCE**
Open Access

www.advancedscience.com

known-operator learning,[17] points in a similar direction and has already shown impressive results by integrating known operators, e.g., subtasks with known analytic solutions in image reconstruction algorithms, into NNs to improve task performance, while preserving the reliability of deterministic methods.[17] However, the decision to integrate priors in known operator learning is a design choice made before the experiments are conducted. SEMPAI's approach is agnostic and decides based on the current task if priors are needed. The regularization by weak constraints in the form of auxiliary losses[1] is particularly interesting as this variant of regularization, in addition to competitive predictive performance for our data, has the benefit of being able to process samples, in which priors are not available or not reliable due to low IQ. SEMPAI has learned the priors during training and implicitly uses them for inference of those cases even without explicit prior computation. A similar concept of regularization, but for dynamical systems, is applied in physics-informed neural networks,[56] which regularize the learning of systems dynamics by known differential equations. Priors are represented by the differential equations that are incorporated into the NN training by losses that use the deviation between predictions made by the NN and those expected following the equations.

Most studies with DL develop/optimize their neural network (NN) architecture for a fixed data representation (DaRe). SEMPAI, however, uses the simultaneous optimization of the DaRe for biological knowledge discovery. Thereby, we showed that most of the investigated learning tasks, as expected, benefit from a higher image resolution. SEMPAI further showed that the muscle periphery is especially important for the assessment of active and passive force measurements or that the distinctive properties of *mdx* dystrophic phenotype are rather learned locally, i.e., at specific locations of the fiber, than globally, i.e., widespread over the whole fiber. However, prediction of mdx by SEMPAI is, to a certain extent, also possible using only global characteristics, which is in concordance with recent literature.[57] The information provided by SEMPAI can be used to guide future experiments and to refine microscopy hardware specifically for a pathology, e.g., by maintaining high resolution in the case of *mdx* or by decreasing resolution in the case of sepsis to increase throughput. Compared to SEMPAI, the recent ground-breaking meta-learning approach of Isensee et al. to the biomedical image segmentation problem[3] is more technically driven by evolving its decision-making around pre-processing and network topology. SEMPAI, however, focuses its decision-making rather on integrating and returning interpretable information regarding prior knowledge and biology.

SEMPAI leverages shared patterns using multi-task learning. The benefit of jointly learning multiple tasks has been shown previously[26,58]; since it allows for a more robust prediction performance even in those tasks for which only a few positive samples are available. Otherwise, with just a small number of examples insufficient for training a high-variance model from scratch, relying on an already established prior would often be the only option for the lab scientist. Notably, joint learning is also interesting for biological reasons, as shown in pan-cancer research,[59] since the highlighting of common patterns between related pathologies might be beneficial in the development of appropriate drugs. In addition to joint learning of multiple tasks by the NN, it was suggested that joint meta-learning, i.e., simultaneous optimiza-

tion of NN architecture and configurations over different tasks might be beneficial.[12] This is explicitly utilized by SEMPAI as well. One main objective of SEMPAI's multi-task learning approach is to create foundation models, i.e., models trained on a multitude of similar tasks, that are then only fine-tuned for a novel task. Foundation models are mostly semi-supervised due to the lack of labels, (pre-)trained on a variety of similar tasks and adapted to the respective application by domain adaptation. For this purpose, further experiments with single muscle fibers and in animal models will be added to the existing database, and, by combining priors and DL, robust foundation models will be generated by SEMPAI. Those foundation models can then be fine-tuned for MPM endoscopy,[60] thereby potentially translating from fundamental research to the clinics.

As one limitation of this study, while intended as a general-purpose tool, SEMPAI was only evaluated for muscle research. In the future, we plan to expand SEMPAI to other organ models, including existing gastroenterological[60-62] and pneumonological[63] MPM databases and respective priors. Further, SEMPAI did not yield a good predictive performance with 3D DL based on the underlying architecture. The phenomenon that DL approaches using lower-dimensional "multi-view" data representations are sometimes superior to DL methods working directly on 3D data is well-known.[64,65] In addition, it is also conceivable that SEMPAI recognized that the data amount was not sufficient for a 3D analysis with significantly more degrees of freedom, and hence regularized itself. However, we believe that further conceptual developments for SEMPAI are required for beneficial use of full 3D information. Another drawback comes from the use of meta-learning. This is very computationally intensive because a large number of models need to be trained. Since we also utilize meta-learning for knowledge discovery, we cannot prune the training as much by, e.g., hyperband pruning[66] or other aggressive pruners. Also, performance estimation strategies from the NAS domain do not seem reliable enough for knowledge discovery. Thus, a single converged run takes between three to four weeks on a system consisting of a Nvidia RTX3090 GPU and an Intel Core-i9 10850k CPU, see Methods. Compared to NAS optimizations, e.g., on ImageNet, which have been performed by industry with supercomputers and large costs, the computation time is moderate with our approach. However, with a further increase of the configuration space, e.g., by more adjustments of the NN architecture as in NAS, we will have to resort to more powerful hardware in the future.

In conclusion, in this work we present SEMPAI, an AI specific for laboratory and basic research. It uses meta-learning for knowledge discovery, allows combining the hypothesis-driven approach of fundamental research with DL, and shows good predictive performance even for small experiments where DL or machine learning in general would not be rationally applicable. It is strongly regularized and prevents overfitting through several external design choices and internal optimization choices. SEMPAI's decision to integrate priors, utilize NN architectures of low capacity, and use low-dimensional DaRe were internal optimization choices for regularization. Its utilization of multi-task learning is an external design choice. We tested this approach with a large exclusive dataset of 3D SHG images of single muscle fibers with a multitude of pathologies and functional properties, which result from over a decade of experiments. The meta-learning on a large database aims to build foundation models for different organs, which could find future application when translating from ex vivo[62] to in vivo experiments[60,61] or from animal models to humans. Both through the systematic analysis of differences and similarities between experiments and pathologies and the adaptation of the method by meta-learning, as well as through the continuous expansion of its database, we expect a continuous self-enhancement of the method.

4. Experimental Section

Selected Studies: *a. A – inflammatory phenotype (sepsis vs. control)*.[30] Sepsis was induced by caecal ligation and puncture (CLP) of 24-week-old mice, and the *extensor digitorum longus* (EDL) muscle was extracted. Maximum isometric tetanic forces were induced in the native whole muscle via needle electrodes (Aurora Scientific) by averaging three consecutive tetanic stimuli (150 Hz stimulation frequency, 200 ms duration, 0.2 ms pulse width, 2 min rest intervals). Thereafter, the dissected and in paraformaldehyd (PFA) fixed muscle tissue was imaged with a voxel size of $0.2 \times 0.2 \times 0.5$ μm, in a field-of-view of 100×100 μm with a stack depth of typically 50 μm. Single fiber biomechanics was assessed using the previously described *MyoRobot* system to measure active force and reconstruct the force-pCa curve. The 3D-SL and myofiber diameter were derived at the beginning of the experiment.

b. B1 & B2 – active force & dystrophic phenotype (mdx vs. WT).[31] The age of the mice was between 13 and 21 weeks for WT and between 27 and 91 weeks for *mdx*. Single muscle fiber segments were manually dissected from the native EDL muscle and clamped into the *MechaMorph* system for subsequent force measurements and SHG imaging. The fiber was adjusted so that its SL was in the range of 2.2 – 3.1 μm as shown by the *MechaMorph* system. Then, force measurements were performed to assess active force parameters (see above). The maximum activation was measured at a pCa of 4.92 in an undiluted highly activating solution (HA, mM: Hepes 30, Mg(OH)$_2$ 6.05, EGTA 30, CaCO$_3$ 29, Na$_2$ATP 8, Na$_2$CP 10, pH 7.2). Specific force, Hill-fit and pCa50 were determined. SHG imaging was performed in two different scenarios (B1 & B2). In B1, a 3D SHG image stack was recorded at each single force recording. In B2, the fiber was only imaged in the relaxed state (pCa 9). Single fibers were z-scanned using a 0.5 μm step size and at a voxel-size of $0.139 \times 0.139 \times 0.500$ μm^3.

c. C – passive force & dystrophic phenotype (mdx vs. WT).[31] The overall procedure was the same as in the active force measurements described above (see B1&B2). However, in this case the *MechaMorph* system was used to access passive force parameters. At each step of force recording, an SHG 3D image stack of the fiber was recorded before proceeding to the next stretch step.

d. D – muscle type (EDL vs. SOL) & dystrophic phenotype (mdx vs WT) in tissue. The investigated mice were 9 months of age. Whole muscle tissue from EDL and diaphragm was fixed in 4% PFA and transferred in PBS on dry-ice for transportation. Each muscle was cut longitudinally at the highest cross-sectional area. Small cryo-cuts of 10 μm were performed and collected on microscope slides. Each slice was further investigated by SHG microscopy. VD, CAS, and SL were derived. In some cases (N = 222), images were recorded from whole muscle tissues. In these cases, single fibers were digitally cropped from the 3D image stacks and afterwards standardized. Force recordings were not performed here.

Label-Free SHG Imaging and Functional Force Measurements: *a. Label-free SHG imaging*. Label-free SHG imaging was performed on an inverse multiphoton microscope (TriMScope, LaVision BioTec, Bielefeld, Germany) with a mode-locked fs-pulsed Ti:Sa laser (Chameleon Vision II, Coherent, Santa Clara, CA, USA). The laser was tuned to a wavelength of 810 nm, generating the second harmonic generation signal at 405 nm. The laser was focused into the sample by a water immersion objective (LD C-Apochromat lens – 40x/1.1/UV–vis-IR/WD 0.62, Carl Zeiss, Jena, Germany), and the generated SHG signal was detected by an ultra-sensitive photo multiplier tube (PMT) (H 7422–40 LV 5 M, Hamamatsu Photonics) in transmission mode to target the SHG of myosin-II.

b. Functional force measurements via the MyoRobot system.[37,38] The MyoRobot was a biomechatronics system for automated assessment of biomechanical active and passive properties as previously described.

c. Functional force measurements via the MechaMorph system.[31] The *MechaMorph* was a custom-engineered device for combined structure–force measurements. A small measurement chamber could be inserted onto the microscope stage below the objective. Single muscle fiber segments could be mounted between a force transducer and a software-controlled voice coil actuator (VCA) that allows the *MechaMorph* to perform subsequent isometric force measurements and structural imaging via SHG microscopy.

Priors: *a. Cosine angle sum (CAS)*. The CAS quantifies the angular deviation of myofibrillar bundles from the main axis.[46] This well-established parameter was deduced from 2D planes of SHG images by a software algorithm (2D-CAS).[46,67,68] The CAS describes disturbances in muscle myofibrillar architecture that have been shown to correlate with muscle weakness.[33] For that an upgraded version for 3D assessment of CAS (3D-CAS) was developed recently.

b. Vernier density (VD). Y-shaped deviations from the regular sarcomere pattern in SHG images are referred to as "verniers". The number of these verniers was then normalized to the fiber area to obtain the VD. Values close to zero represent fibers, where all myofibrils were perfectly in register, while larger values indicate deviations. The VD can either be generated manually or by a custom-designed software tool.[68]

c. Sarcomere length (SL). With the software tools for *MechaMorph* and *MyoRobot*, the SL was recorded live.

d. Smart Cross-Sectional Area (CSA) computation. In the current study, a new method for quantifying the CSA of single muscle fibers is reported, which had been developed for a standardized solution of the CSA in all image data sets. First, a binarization of the images was performed by a simple Otsu threshold on the images. An oriented bounding box algorithm[69] was applied to the binarized fiber to orient the fiber vertically. The top and bottom 10 slices were excluded from quantification. Then, three algorithms were combined with each other, and an outlier detection was applied to increase the stability of the method.

I. Algorithm 1 – exact counting: Since the binarized fiber was now arranged from top to bottom, morphological operations 2D opening and closing were applied to each slice to close holes and obtain a compact segmentation. After application, the number of pixels in each slice was counted and averaged.

II. Algorithm 2 – principal component-based: Instead of morphological operations, a 2D principal component analysis (PCA) of scikit-learn was applied and the obtained maximum and minimum radii were used to determine the area of an ellipse for each slice. The results were averaged over the slices.

III. Algorithm 3 – elliptic envelope-based: Instead of morphological operations, an elliptic envelope (EE) was calculated with a contamination of 0.2. The area of the EE was calculated for each slice and the results were averaged across slices.

The mean results of two algorithms, which show higher concordance, were used. The averaging and outlier removal compensates for potential weaknesses of the algorithms due to varying IQ. The results agreed well with visual assessment.

Implementation of cross-study standardization and data split: The pipeline was written in Python (v3.7.7). For studies with low SNR, a median filter of size 1 μm was applied. An intensity threshold for the background by Otsu's thresholding was defined. Then, voxels with intensities below this threshold intensity were set to 0 (background). The contrast enhancement algorithm MCLAHE[47] was applied with adaptive histogram range. The registration toolbox Elastix[70] was used to register the muscle fibers to a reference fiber, which exhibits a canonical structure and perfectly vertical orientation. A rigid multi-scale Euler registration with 600 iterations was used, automatic scale estimation, center of gravity initialization, 32 bins, 6 scales, and grid-adaptive step size. The transformation was then also applied to the non-enhanced fiber. Each standardized fiber was normalized to a sample-wise standard score. Force measurements were extracted

ADVANCED
SCIENCE NEWS

www.advancedsciencenews.com

ADVANCED
SCIENCE
Open Access

www.advancedscience.com

directly from the TDMS curves coming from the instruments, entered the data frame and normalized by the CSA of the associated fiber. The standardization pipeline was highly automated, and the steps were documented by an automatically generated SEMPAI labbook to identify and minimize errors associated with standardization or data management.

For data splitting in train (2/4), dev (1/4) and test (1/4) set, the data were both stratified and grouped. The stratification was needed to had sufficient data with a certain label in all sets. Continuous functional labels were median-dichotomized into "high" and "low" values, e.g., specific force "high" for stratification. However, those dichotomized labels were only used for stratification and not as a learning task. This stratification also ensures that class distributions were balanced over train, dev, and test set. The labels were grouped according to muscle bundle, single fibers from one bundle were therefore, not split between train, dev and test set, preventing information leakage.

Implementation of SEMPAI configuration-space and self-enhancement. SEMPAI was implemented in Python (v3.8.1), its NN parts in PyTorch (v1.11, CUDA v11.3). For meta-learning, the multi-objective optimization algorithm NSGA-II[29] from the Optuna[71] package was leveraged with population size of 50, without mutation probability, with a crossover probability of 0.9, swapping probability of 0.5, and a fixed seed of 42.

The losses of labels and priors were weighed against each other by uncertainty weighing.[28] For this purpose, additional learnable parameters were introduced, that weigh the losses against each other. The loss is, therefore, determined by: $\mathcal{L} = \sum_i (\frac{\mathcal{L}_{L,i}}{\sigma_{L,i}^2} + \log\sigma_{L,i}) + \sum_j (\frac{\mathcal{L}_{P,j}}{\sigma_{P,j}^2} + \log\sigma_{P,j})$ with labels i of set L and priors j of set P, and the learnable uncertainties associated with each label $\sigma_{L,i}$ and prior $\sigma_{P,j}$. For the *2.5D* DaRes, three 2D slices of the 3D images were fed in three channels of a 2D EfficientNet. The center slice of the cropped bounding box was used and two further peripheral slices, whose distance from the center slice was optimized by SEMPAI. For NN with branches, i.e., SEMPAI *Branches* and *AuxLosses@Branches*, a wrapper was built for the respective NN to introduce the priors in the fully connected layers.

For AutoML based on priors, i.e., SEMPAI *PriorsOnly*, the Tree-based Pipeline Optimization Tool (TPOT)[72] was employed. This algorithm combines identification of feature selection and suitable classifiers or regressors with Pareto optimization. 250 generations was used, a population size of 200, and grouping of the fibers. The combined train & dev set was forwarded to TPOT for training, and the internal cross-validation (CV) was set to two-fold to have a comparable data split ratio to the other components of SEMPAI. TPOT was restricted to methods with class probability output. The performance metric, e.g., *AUC* or R^2, of the internal CV was reported to SEMPAI and evaluated as a meta-loss, i.e., the model selected by SEMPAI can be a prior-only model based on AutoML.

The *total meta-loss* was a weighted sum of each label. The labels were weighted as a trade-off between sample size and importance of task, accordingly we set weights w = [*mdx*: 1.0, sepsis: 1.0, muscle type: 0.5, active force: 1.0, active force/pCa: 1.0, passive force: 1.0, pCa50 = 1.0]. In the trade-off between exploration and exploitation, multi-objective optimization algorithms were lending toward exploration as the performance for different tasks must be optimized. Thus, the configuration space was sufficiently sampled although very unpromising regions of configuration space trials were still under-sampled. Selecting a criterion time for early termination of the trials was not trivial for multi-objective optimization trials. Therefore, a very non-conservative criterion was selected. Accordingly, SEMPAI does not compare trials for termination (and save computation time) as in more modern methods like Hyperband pruning.[66] The *total meta-loss* was smoothed by computation of the moving average of the last 10 epochs. A trial was terminated when the *total meta-loss* did not decrease for 50 subsequent epochs. The early stopping criterion was set active after the initial 75 epochs, resulting in at least 125 epochs performed per trial. The lowest meta-loss for each respective task was used to select the respective model for the task. For tasks with scarce data (pCa50, passive force), however, the *total meta-loss* was used for model selection.

SEMPAI offers sample-level and model-level decision explanations. For a more detailed explanation of the sample-level explanation, the example in Figure 3 was used. The sample-level explanation uses Deep

SHAP[15]. The explanations were adapted to the DaRe and prior integration variants. For the prior integration variants with branches, where the priors were fed as branches into the fully connected layer of the respective NN architecture, the importance of these priors was calculated simultaneously to the importance of the voxels of the image. In Figure 3, it can be seen that SEMPAI uses the twisted image regions and the vernier density to correctly classify this sample as dystrophic. In addition, a model-level explanation was provided by SEMPAI: to provide more insights about preferable individual configurations for each target label, SHapley Additive exPlanations (SHAP)[15] values were computed. For this purpose, a random forest was trained to predict performance metrics of the dev set based on the configuration space. The SHAP Tree Explainer[35] was utilized, which was explicitly designed for tree-based algorithms like the fitted model. The stability of the results by fitting multiple forests with different random initializations was verified and ensemble sizes (i.e., number of trees). Manually inspecting each resulting plot of two representative labels (*mdx* and active force/pCa) gave rise to the same interpretation.

For an application of SEMPAI to other data bases, e.g., other organs, corresponding handcrafted features must be provided as priors, or their computation must be integrated into SEMPAI. Ideally, these were known biomarkers or good hypotheses. A dataframe must be created for train, dev, and test set, containing the labels, the priors, and the paths to the standardized images. Then the meta-parameter space must be defined for SEMPAI, i.e., which configurations regarding DaRe, prior integration, and NN architecture will be tested and optimized.

Rationale for Standardization and Configuration Space: a. Standardization: Standardization was intended to minimize technical variance, which is usually present in biomarker research.[73] This technical variance can even lead to wrong conclusions of an AI system.[74,75] To reduce the impact of technical variance, The image was slightly denoised and resampled to uniform isotropic voxel size. Cropping reduces the dimensionality of the images, and DL can focus only on relevant regions. The alignment of fibers via registration helps to minimize the bounding box and can increase the convergence of the learning process, because CNNs, such as the employed EfficientNet, are not rotation invariant.

b. Configuration space:

- The benefit of contrast enhancement for visual recognizability of structures was undisputed. However, it was not yet understood if this enhancement adds value for training an AI. Therefore, SEMPAI validates this explicitly and exemplarily for the MCLAHE[47] algorithm.
- Random erasing[48] regularizes the learning process by enforcing the use of multiple image regions for inference, theoretically resulting in a more robust prediction. Random erasing can enforce regularization since it prevents the model from overfitting specific image regions. Thus, the model needs to use several different image regions and can thus become stable and less prone to overfitting to localized noise or other undesired effects. In the case of localized biologically-relevant image properties, however, deleting this location naturally leads to a mis-evaluation of the image and a decrease in predictive performance. We thus use random sampling as a measure for the importance of localized image features.
- Downsampling and multi-view representations may support learning by minimizing overfitting. It was scientifically interesting to understand the importance of resolution for learning phenotypes and function, since microscopy research targets finer resolution (lower pixel size), often at the expense of reduced throughput. SEMPAI's decision w.r.t. down-sampling to elaborate how important the image resolution was for a given learning task was interpreted. In analogy, whether to use 3D data directly for learning was evaluated, or to draw representative 2D slices. Whether using lower dimensional DaRe as NN input via downsampling (reduced voxel size) and sub-sampling (2.5D vs 3D), was tested improves convergence. The benefit of dimensionality reduction in DL was controversial.[76,77] Choosing the spacing of the representative slices was also of biological interest. It allows interpretation of where relevant information was located in 3D, i.e., by interpretably sub-sampling a lower-dimensional DaRe from a higher dimensional

volume. By this, the origin of the biological information can be narrowed down.

- To test the importance of priors, several prior integration methods was used. Besides both extremes, *NoPriors* and the *PriorsOnly*, priors as auxiliary tasks was used, as branches or as a combination of the latter two, to define a scale of prior integration from "weak to strong". By defining the priors as auxiliary tasks, they were predicted simultaneously to the labels. Thus, the filter kernels of the NN evolve to predict these auxiliary tasks as well. By using these priors as auxiliary tasks, the network can leverage domain knowledge to learn better representations of the data. Since the prior was only indirectly available for learning a label, it as weak prior integration was considered. With the branches approach, the priors were passed on directly to the fully connected layers of the NN, i.e., theoretically, the NN can completely dispense with the additional image information, which was why it as strong prior integration was considered. By adding a prior, i.e., handcrafted feature, branch to the fully connected layer, the network can learn to combine the learned features of the CNN with the priors, which can potentially improve the accuracy and generalization of the model. This approach can be particularly useful in scenarios where the input data was noisy or incomplete, and the handcrafted features can provide additional information to the network. Multi-branch approaches have also recently been shown to have positive convergence properties[78] for learning. The combination of both methods as an even stronger prior integration was defined. Finally, the use of priors with AutoML, i.e., without images and DL, was defined as the "maximum" of the prior integration scale. Such feature-based ML approaches can occasionally outperform DL.[79] In the optimization of SEMPAI, the added value of the priors for the learning process was evaluated. If models with the hypothesis-driven priors were superior to models without, or if a prior-only model shows the same performance as the best DL model, the hypothesis that the prior describes the state of the label well can be considered true. The researcher can thus test hypotheses and these were verified by SEMPAI and, in the case of models with DL, also refined. The biological information of the prior knowledge was evaluated.

- Further adaptations: NN-specific parameters were more technical and less interpretable but need to be adapted to prior integration and DaRe at hand to achieve a global optimum. The NN capacity was adjusted, as it must be adapted to the available amount of data and the complexity of the learning task. Also, further NN properties like batch size, learning rate, momentum, and optimizer must be fine-tuned. Gradient clipping, i.e., restricting the gradients, had been theoretically shown to accelerate convergence[80] and its benefit was evaluated. Also, the sampling of the data can be modified by imbalance sampling. The employed augmentation uses rotations, shifts, and additive noise patterns, which were identified as variations in the data after inspection of the images by domain experts. Thus, this step can also be interpreted as prior knowledge integration. Augmentation introduces invariance toward the applied modifications to the learning process.

Computation Details: SEMPAI computed 19 days on a workstation equipped with NVMe SSD, Nvidia RTX 3090 GPU and Intel Core-i9 10850k CPU (10 cores of 3.6 GHz), resulting in a total of 1,500 evaluated trials. To decrease the computational cost for evaluating 2D configurations, the slices were loaded by reading parts of the memory-mapped 3D volume. For (3D) augmentations, some operations employ TorchIO,[81] and where possible, augmentations were computed on the GPU. Automated mixed precision (AMP) of PyTorch was used in addition to multiple workers and pinned memory. To be able to use sufficiently large batches for 3D data, SEMPAI utilizes gradient accumulation.

Supporting Information

Supporting Information is available from the Wiley Online Library or from the author.

Acknowledgements

This project had received funding from the European Union's Horizon Marie Skłodowska-Curie Action under grant agreements 101103200 (project MICS to LK) and 812772 (project Phys2BioMed to WHG). The project was further supported by the 2021 Emerging Talents Initiative of the Friedrich-Alexander University (to LK), as well as the DFG (German Research Foundation) projects TRR225-Z02 -Biofabrication (326998133 to OF) and TRR241-C01 (to SS).

Open access funding enabled and organized by Projekt DEAL.

Conflict of Interest

The authors declare no conflict of interest.

Author Contributions

S.N., D.S., C.G., P.R., S.S. conducted the original experiments. L.K., A.M., P.R. curated the data. A.M. designed the research. A.M. conceptualized and developed the method and the software. A.M., L.K. wrote the initial draft; all authors revised the draft. A.M., L.K., P.R., M.H., S.L., D.N., R.H., W.H.G., O.F. conducted the literature review. A.K.M., W.H.G., L.K., O.F. supervised the work. L.K., P.R. evaluated the integrity of multi-study data, labels, and priors. O.T., S.L., F.D. evaluated the integrity of the method. O.T., F.D., S.L. reviewed the method development and software. D.N, R.H., W.H.G., O.F. contributed to the interpretation of the results.

Data Availability Statement

The complete code is provided on Google Colab under https://colab. research.google.com/drive/18foBUuWKZEVPNuvYaRfmczlaWaaMg6mP All dependencies are pre-installed and standardized images are provided for rapid execution of the SEMPAI software. The code to standardize the images is also being provided. Finally, the log of a second non-converged SEMPAI run for another data split is also made available along with analysis scripts, to enable a deeper understanding.

Keywords

deep learning, explainable artificial intelligence, meta-learning, multiphoton microscopy, muscle research, prior information integration, scientific machine learning

Received: October 28, 2022
Revised: June 30, 2023
Published online:

[1] B. S. Doken, T. Zhuang, D. Wingerter, M. Gidwani, N. Mistry, L. Ladic, A. Kamen, M. E. Abazeed, *The Lancet Digital Health* **2019**, *1*, e136.

[2] S. M. Humphries, A. M. Notary, J. P. Centeno, M. J. Strand, J. D. Crapo, E. K. Silverman, D. A. Lynch, *Radiology* **2020**, *294*, 434.

[3] F. Isensee, P. F. Jaeger, S. A. A. Kohl, J. Petersen, K. H. Maier-Hein, *Nat. Methods* **2021**, *18*, 203.

[4] G. Litjens, T. Kooi, B. E. Bejnordi, A. A. A. Setio, F. Ciompi, M. Ghafoorian, J. A. W. M. Van Der Laak, B. Van Ginneken, C. I. Sánchez, *Med. Image Anal.* **2017**, *42*, 60.

[5] G. Aresta, T. Araújo, S. Kwok, S. S. Chennamsetty, M. Safwan, V. Alex, B. Marami, M. Prastawa, M. Chan, M. Donovan, G. Fernandez, J. Zeineh, M. Kohl, C. Walz, F. Ludwig, S. Braunewell, M. Baust, Q. D. Vu, M. N. N. To, E. Kim, J. T. Kwak, S. Galal, V. Sanchez-Freire, N. Brancati, M. Frucci, D. Riccio, Y. Wang, L. Sun, K. Ma, J. Fang, et al., *Med. Image Anal.* **2019**, *56*, 122.

**ADVANCED
SCIENCE NEWS**

www.advancedsciencenews.com

**ADVANCED
SCIENCE**
Open Access

www.advancedscience.com

[6] A. Esteva, B. Kuprel, R. A. Novoa, J. Ko, S. M. Swetter, H. M. Blau, S. Thrun, *Nature* **2017**, *542*, 115.

[7] O. Ronneberger, P. Fischer, T. Brox, in *International Conference on Medical Image Computing and Computer-Assisted Intervention*, Springer, **2015**, 234.

[8] Y. F. Cheng, M. Strachan, Z. Weiss, M. Deb, D. Carone, V. Ganapati, *Opt. Express* **2019**, *27*, 644.

[9] E. Nehme, L. E. Weiss, T. Michaeli, Y. Shechtman, *Optica* **2018**, *5*, 458.

[10] M. Weigert, U. Schmidt, T. Boothe, A. Müller, A. Dibrov, A. Jain, B. Wilhelm, D. Schmidt, C. Broaddus, S. Culley, M. Rocha-Martins, F. Segovia-Miranda, C. Norden, R. Henriques, M. Zerial, M. Solimena, J. Rink, P. Tomancak, L. Royer, F. Jug, E. W. Myers, *Nat. Methods* **2018**, *15*, 1090.

[11] T. Elsken, J. H. Metzen, F. Hutter, *J. Mach. Learn. Res.* **2019**, *20*, 1997.

[12] K. O. Stanley, J. Clune, J. Lehman, R. Miikkulainen, *Nat. Mach. Intell.* **2019**, *1*, 24.

[13] Y. Liu, W. Zhao, L. Liu, D. Li, S. Tong, C. L. P. Chen, *IEEE Transactions on Neural Networks and Learning Systems* **2021**, *34*, 2732.

[14] S. Jiao, Y. Gao, J. Feng, T. Lei, X. Yuan, *Opt. Express* **2020**, *28*, 3717.

[15] S. M. Lundberg, S.-I. Lee, *Adv. Neural Inf. Process Syst.* **2017**, *30*.

[16] C. Rudin, *Nat. Mach. Intell.* **2019**, *1*, 206.

[17] A. K. Maier, C. Syben, B. Stimpel, T. Würfl, M. Hoffmann, F. Schebesch, W. Fu, L. Mill, L. Kling, S. Christiansen, *Nat. Mach. Intell.* **2019**, *1*, 373.

[18] L. Tian, J. Wang, L. Waller, *Opt. Lett.* **2014**, *39*, 1326.

[19] R. Horstmeyer, C. Yang, *Opt. Express* **2014**, *22*, 338.

[20] M. E. Kandel, Y. R. He, Y. J. Lee, T. H.-Y. Chen, K. M. Sullivan, O. Aydin, M. T. A. Saif, H. Kong, N. Sobh, G. Popescu, *Nat. Comm.* **2020**, *11*, 6256.

[21] C. L. Cooke, et al., Proceedings of the IEEE/CVF International Conference on Computer Vision **2021**, 3803.

[22] M. Moor, O. Banerjee, Z. S. H. Abad, H. M. Krumholz, J. Leskovec, E. J. Topol, P. Rajpurkar, *Nature* **2023**, *616*, 259.

[23] R. Bommasani, et al., arXiv preprint arXiv:2108.07258, **2021**.

[24] L. Parmentier, O. Nicol, L. Jourdan, M.-E. Kessaci, in *2019 IEEE 31st International Conference on Tools with Artificial Intelligence (ICTAI)*, IEEE, **2019**, 471.

[25] M. Tan, Q. Le, in *International Conference on Machine Learning*, PMLR, **2019**, 6105.

[26] R. M. Caruana, *Multitask Learning* **1997**, *28*, 41.

[27] G. M. Van De Ven, H. T. Siegelmann, A S. Tolias, *Nat. Comm.* **2020**, *11*, 4069.

[28] A. Kendall, Y. Gal, R. Cipolla, in Proceedings of the IEEE Conference on Computer Vision and Pattern Recognition **2018**, 7482.

[29] K. Deb, A. Pratap, S. Agarwal, T. Meyarivan, *IEEE Trans. Evol. Comput.* **2002**, *6*, 182.

[30] C. Goossens, R. Weckx, S. Derde, L. Van Helleputte, D. Schneidereit, M. Haug, B. Reischl, O. Friedrich, L. Van Den Bosch, G. Van Den Berghe, L. Langouche, *J. Cachexia Sarcopeni.* **2021**, *12*, 443.

[31] D. Schneidereit, S. Nübler, G. Prölß, B. Reischl, S. Schürmann, O. J. Müller, O. Friedrich, *Light Sci. Appl.* **2018**, *7*, 79.

[32] P. Ritter, *International journal of molecular sciences* **2022**, *23*, 10841.

[33] S. Diermeier, et al., *Stem Cells Int* **2017**, *7*.

[34] S. Diermeier, M. Haug, B. Reischl, A. Buttgereit, S. Schürmann, M. Spörrer, W. H. Goldmann, B. Fabry, F. Elhimine, R. Stehle, G. Pfitzer, L. Winter, C. Clemen, R. Schröder, O. Friedrich, *Biophys. J.* **2016**, *110*, 303a.

[35] S. M. Lundberg, G. Erion, H. Chen, A. Degrave, J. M. Prutkin, B. Nair, R. Katz, J. Himmelfarb, N. Bansal, S.-I. Lee, *Nat. Mach. Intell.* **2020**, *2*, 56.

[36] A. E. H. Emery, *Clin. Genet.* **1983**, *23*, 198.

[37] M. Haug, C. Meyer, B. Reischl, G. Prölß, S. Nübler, S. Schürmann, D. Schneidereit, M. Heckel, T. Pöschel, S. J. Rupitsch, O. Friedrich, *Biosens. Bioelectron.* **2019**, *138*, 111284.

[38] M. Haug, B. Reischl, G. Prölß, C. Pollmann, T. Buckert, C. Keidel, S. Schürmann, M. Hock, S. Rupitsch, M. Heckel, T. Pöschel, T. Scheibel, C. Haynl, L. Kiriaev, S. Head, O. Friedrich, *Biosens. Bioelectron.* **2018**, *102*, 589.

[39] O. Friedrich, M. Haug, B. Reischl, G. Prölß, L. Kiriaev, S. I. Head, M. B. Reid, *The International Journal of Biochemistry, Cell Biology* **2019**, *114*, 105563.

[40] D. G. Moisescu, *Nature* **1976**, *262*, 610.

[41] A. Fischmann, P. Hafner, M. Gloor, M. Schmid, A. Klein, U. Pohlman, T. Waltz, R. Gonzalez, T. Haas, O. Bieri, D. Fischer, *J. Neurol.* **2013**, *260*, 969.

[42] M. Liu, N. Chino, T. Ishihara, *Arch. Phys. Med. Rehabil.* **1993**, *74*, 507.

[43] M. Jansen, N. Van Alfen, M. W. G. Nijhuis Van Der Sanden, J. P. Van Dijk, S. Pillen, I. J. M. De Groot, *Neuromuscular Disord.* **2012**, *22*, 306.

[44] N. Brouilly, C. Lecroisey, E. Martin, L. Pierson, M.-C. Mariol, H. Qadota, M. Labouesse, N. Streichenberger, N. Mounier, K. Gieseler, *Hum. Mol. Genet.* **2015**, *24*, 6428.

[45] M. Dubreuil, F. Tissier, L. Le Roy, J.-P. Pennec, S. Rivet, M.-A. Giroux-Metges, Y. Le Grand, *Biomed. Opt. Express* **2018**, *9*, 6350.

[46] C. S. Garbe, A. Buttgereit, S. Schurmann, O. Friedrich, *IEEE Trans. Biomed. Eng.* **2012**, *59*, 39.

[47] V. Stimper, S. Bauer, R. Ernstorfer, B. Scholkopf, R. P. Xian, *IEEE Access* **2019**, *7*, 165437.

[48] Z. Zhong, L. Zheng, G. Kang, S. Li, Y. Yang, in *Proceedings of the AAAI Conference on Artificial Intelligence*, **2020**, *34*, 13001.

[49] G. E. Hinton, N. Srivastava, A. Krizhevsky, I. Sutskever, R. R. Salakhutdinov, arXiv preprint arXiv:1207.0580, **2012**.

[50] C. Ding, H. Peng, *J. Bioinf. Comput. Biol.* **2005**, *3*, 185.

[51] A. Khamparia, A. Singh, D. Anand, D. Gupta, A. Khanna, N. Arun Kumar, J. Tan, *Neural Comput. Appl.* **2020**, *32*, 11083.

[52] A. H. Liao, J.-R. Chen, S.-H. Liu, C.-H. Lu, C.-W. Lin, J.-Y. Shieh, W.-C. Weng, P.-H. Tsui, *Diagnostics* **2021**, *11*, 963.

[53] A. Mankodi, W. Kovacs, G. Norato, N. Hsieh, W. P. Bandettini, C. A. Bishop, H. Shimellis, R. D. Newbould, E. Kim, K. H. Fischbeck, A. E. Arai, J. Yao, *Ann. Clin. Transl. Neurol.* **2017**, *4*, 655.

[54] J. Cai, F. Xing, A. Batra, F. Liu, G. A. Walter, K. Vandenborne, L. Yang, *Pattern Recogn.* **2019**, *86*, 368.

[55] M. Yang, et al., *BMC Neurol.* **2021**, *21*, 1.

[56] G. E. Karniadakis, I. G. Kevrekidis, L. Lu, P. Perdikaris, S. Wang, L. Yang, *Nat. Rev. Phys.* **2021**, *3*, 422.

[57] L. Kiriaev, S. Kueh, J. W. Morley, P. J. Houweling, S. Chan, K. N. North, S. I. Head, *Am. J. Physiol.: Cell Physiol.* **2021**, *321*, C704.

[58] E. Meyerson, R. Miikkulainen, in *International Conference on Machine Learning*, PMLR, **2018**, 3511.

[59] J. N. Weinstein, E. A. Collisson, G. B. Mills, K. R. M. Shaw, B. A. Ozenberger, K. Ellrott, I. Shmulevich, C. Sander, J. M. Stuart, *Nat. Genet.* **2013**, *45*, 1113.

[60] A. Dilipkumar, A. Al-Shemmary, L. Kreiß, K. Cvecek, B. Carlé, F. Knieling, J. Gonzales Menezes, O.-M. Thoma, M. Schmidt, M. F. Neurath, M. Waldner, O. Friedrich, S. Schürmann, *Adv. Sci.* **2019**, *6*, 1801735.

[61] L. Kreiß, O.-M. Thoma, A. Dilipkumar, B. Carlé, P. Longequeue, T. Kunert, T. Rath, K. Hildner, C. Neufert, M. Vieth, M. F. Neurath, O. Friedrich, S. Schürmann, M. J. Waldner, *Gastroenterology* **2020**, *159*, 832.

[62] L. Kreiss, O.-M. Thoma, S. Lemire, K. Lechner, B. Carlé, A. Dilipkumar, T. Kunert, K. Scheibe, C. Heichler, A.-L. Merten, B. Weigmann, C. Neufert, K. Hildner, M. Vieth, M. F. Neurath, O. Friedrich, S. Schürmann, M. J. Waldner, *Inflamm. Bowel Dis.* **2022**, *28*, 1637.

[63] L. Kreiss, I. Ganzleben, A. Mühlberg, P. Ritter, D. Schneidereit, C. Becker, M. F. Neurath, O. Friedrich, S. Schürmann, M. Waldner, *J. Biophotonics* **2022**, *15*, 202200073.

[64] E. Ahmed, et al., arXiv Preprint, 1808.01462, **2018**.

[65] F. Denzinger, et al., *International Conference on Medical Image Computing and Computer-Assisted Intervention*, Springer, **2020**, 45.

[66] L. Li, K. Jamieson, G. DeSalvo, A. Rostamizadeh, A. Talwalkar, *J. Mach. Lear. Res.* **2017**, *18*, 6765.

[67] O. Friedrich, M. Both, C. Weber, S. Schürmann, M. D. H. Teichmann, F. Von Wegner, R. H. A. Fink, M. Vogel, J. S. Chamberlain, C. Garbe, *Biophys. J.* **2010**, *98*, 606.

[68] A. Buttgereit, C. Weber, C. S. Garbe, O. Friedrich, *J. Pathol.* **2013**, *229*, 477.

[69] S. A. Gottschalk, *Collision queries using oriented bounding boxes*, The University of North Carolina at Chapel Hill, **2000**.

[70] S. Klein, M. Staring, K. Murphy, M. A. Viergever, J. Pluim, *IEEE Transactions on Medical Imaging* **2009**, *29*, 196.

[71] T. Akiba, S. Sano, T. Yanase, T. Ohta, M. Koyama, in Proceedings of the 25th ACM SIGKDD International Conference on Knowledge Discovery, Data Mining **2019**, 2623.

[72] R. S. Olson, J. H. Moore, in *Automated Machine Learning*, Springer, **2019**, 151.

[73] J. T. Leek, R. B. Scharpf, H. C. Bravo, D. Simcha, B. Langmead, W. E. Johnson, D. Geman, K. Baggerly, R. A. Irizarry, *Nat. Rev. Genet.* **2010**, *11*, 733.

[74] S. Lapuschkin, S. Wäldchen, A. Binder, G. Montavon, W. Samek, K.-R. Müller, *Nat. Comm.* **2019**, *10*, 1096.

[75] S. Langer, O. Taubmann, F. Denzinger, A. Maier, A. Mühlberg, in *Mitigating Unknown Bias in Deep Learning-based Assessment of CT Images DeepTechnome* (Eds: T. M. Deserno, H. Handels, A. Maier, K. Maier-Hein, C. Palm, T. Tolxdorff), Bildverarbeitung für die Medizin (BVM) 2023. Informatik aktuell. Springer, Wiesbaden, **2023**.

[76] T. Poggio, H. Mhaskar, L. Rosasco, B. Miranda, Q. Liao, *Int. J. Autom. Comput.* **2017**, *14*, 503.

[77] P. Pope, C. Zhu, A. Abdelkader, M. Goldblum, T. Goldstein, ArXiv Preprint, 2104.08894, **2021**.

[78] H. Zhang, J. Shao, R. Salakhutdinov, in *The 22nd International Conference on Artificial Intelligence and Statistics*, PMLR, **2019**, pp. 1099-1109.

[79] A. Mühlberg, R. Kärgel, A. Katzmann, F. Durlak, P. E. Allard, J. B. Faivre, M. Sühling, M. Rémy-Jardin, O. Taubmann, *Medical physics* **2021**, *48*, 5179.

[80] J. Zhang, T. He, S. Sra, A. Jadbabaie, ArXiv Preprint, 1905.11881, **2019**.

[81] F. Pérez-García, R. Sparks, S. Ourselin, *Computer Methods and Programs in Biomedicine* **2021**, *208*, 106236.